GIRAFFE REFLECTIONS

Text by **DALE PETERSON**

Photographs by **KARL AMMANN**

University of California Press

Berkeley Los Angeles London

The publisher gratefully acknowledges the generous support of the General Endowment Fund of the University of California Press Foundation.

University of California Press, one of the most distinguished university presses in the United States, enriches lives around the world by advancing scholarship in the humanities, social sciences, and natural sciences. Its activities are supported by the UC Press Foundation and by philanthropic contributions from individuals and institutions. For more information, visit www.ucpress.edu.

University of California Press
Berkeley and Los Angeles, California

University of California Press, Ltd.
London, England

Library of Congress Cataloging-in-Publication Data

Peterson, Dale.
Giraffe reflections / text by Dale Peterson ; photographs by Karl Ammann.
pages cm
Includes bibliographical references and index.
ISBN 978-0-520-26685-8 (cloth : alk. paper)
1. Giraffe. 2. Giraffe—Pictorial works. I. Ammann, Karl. II. Title.
QL737.U56P48 2013
599.638—dc23

2012038611

Manufactured in China

22 21 20 19 18 17 16 15 14 13
10 9 8 7 6 5 4 3 2 1

The paper used in this publication meets the minimum requirements of ANSI/NISO Z39.48–1992 (R 2002) (*Permanence of Paper*).

To the memory of Graham Grindlay (1978–2011)

CONTENTS

SPIRITS

IT WAS STILL DARK when Karl and I left camp. On the way to where we thought the giraffes might be, we passed through a feeding group of Thomson's gazelles. Illuminated starkly by our headlights, they looked like precious tchotchkes: delicate little legs, prancing style, nervously tic-tocking tails.

Fifteen or twenty minutes later, we surprised three giraffes lying down in the grass and looking dazed, as if they had just woken up after a long and satisfying night's sleep. They were emerging from the darkness, bathed faintly in the light-speckling dawn, and all we saw at first was what appeared to be three swaying trees with heads on top. They became alert as we drove closer. They were in a small meadow, edged, protected perhaps, on two sides by a bit of dark and thickety bush, and I imagined the spot as a comfortable bedroom for giraffes.

Karl took some photographs, but instead of waiting patiently for the sun to rise and cast some interesting morning light on the sleeping beauties, he kept driving around, looking for a better angle, taking one or two quick shots with the engine off, then starting the car, moving to a new position. As he worked, he commented on his photography, the animals, the light. "Yeah, it's the nice type of light which says they're just getting up," he said. But the moment was quickly gone. Soon the light had turned slightly harder, and the three lying-down giraffes were laboriously standing up. Then, slowly, they sauntered away. The sun rose and turned into a seething red ball at the horizon, and so the day began.

This happened in southwestern Kenya, in the Masai Mara: a rare place where the modern catastrophe has not yet fully dawned, where, in the fading darkness, it is momentarily possible to believe you have reached the fragile beginning of time.

In the Mara, we saw giraffes singly, doubly. We saw them in groups of three or four or a dozen or more. One time we emerged from a hiding place in the thickets and discovered a group of eighteen. They were Masai giraffes, of course, patterned with brown and splotchy spots. One looked as if she had been made entirely of cream and then one day had been struck forcibly by a mad flock of brown-sugared birds.

One was lying down, the rest standing, all with their ears flickering and their tufted tails desultorily flicking back and forth. They stared. We stared. They stared and chewed their cuds. We stared and took pictures. They stared and then looked at each other. We stared, took pictures, and then Karl started up the car to move closer and get a better position. Several minutes later, I saw a subtle emergence of giraffe consensus. One turned, another turned, a third turned. Soon a half dozen had turned, and then they were all ambling undulously along, stately and elegant. Karl (working on his lenses, muttering to himself): "Try to do a very wide angle once, take them all in."

—

Later, in northern Kenya's Samburu National Reserve, we came upon a group of six reticulated giraffes (whose markings look like brown plates caught in nets of pale hemp) browsing in a nice pocket of trees and bushes. Four of them looked young, one of them very young. They were spread out at first and chewing at the trees and bushes, but eventually they moved out of the pocket and began slowly, patiently ambling in the same uphill direction. They seemed so finely built, so delicate, and they gradually arranged themselves, as they walked, into single file, the four youngsters in the middle, the big adult male at the rear, the adult female at the head.

Then, for no good reason, they stopped and gathered to think about things, or so it seemed. They stood still. They looked in several directions. We saw, then, two more giraffes at some distance behind and moving uphill in their direction. The stragglers looked like adolescents or, possibly, full adults. But everything happened very slowly, and Karl and I remained in the car and then settled into another experience of time, where we were immersed in the sweet smell of dry grass and cooled by a dry wind blowing through the windows, heated by a slanting late afternoon sun, fitfully distracted by the buzzing of a fly. I thought: blond savanna, brown bushes: fitting colors for a giraffe. Meanwhile, the tall female outside our vehicle stared at us for a very long time, then began eating a small cache of green leaves edging a brown, thorny bush, while the two stragglers behind her slowly, slowly began to catch up. Now there were eight in the group, pausing, looking, pausing, browsing, pausing.

A big male had a half dozen red-billed oxpeckers lined up on his back, picking away at a feast of ticks. Karl: "That's quite a lineup. Must be something tasty."

We followed them all slowly, the car grinding away in its lowest gear and struggling heroically over a rough surface of bumps and holes, following the giraffes as they slowly continued uphill, pausing opportunistically at each greenish-brown thornbush. They took bites, too, from the occasional high acacia tree, each filled with a hundred weaverbird nests that dangled like Chinese lanterns. I gazed away momentarily, looking out across a spectacular vista of sun-yellowed plains dropping down to a green-lined river. Then I returned to the giraffes and was suddenly amazed at how narrow their necks are, ribbony even, yet very flexible and immensely strong.

—

In the Namibian desert, at a place called Twyfelfontein, we found giraffes in their most ancient and ethereal form: wispy, rising representations carved into rock by Bushman artists who lived a few or several thousand years ago.

Twyfelfontein. A recent name, Afrikaans in origin, it describes the wistful hope a white farmer formed for this spare spot in the sparse desert. The name translates into English as Doubtful Spring.

The Bushmen camped in a small plateau or terrace just above the doubtful spring, and their camp was a gathering place, a passing refuge in the hard life of hunting and gathering. They were protected by a high cliff and mountain behind them, while before them lay the flat and splendid valley consisting mainly of rust-red stone and sand, which is spotted, after the rains, by the green of small thorn trees and scrub. The valley is surrounded by flattened, red-rocked mountains. The red rocks are Etjo sandstone, consisting of alluvial conglomerates and eolian sandstone—stone, that is, formed from sand that has been sifted by the wind and is thus fine grained and capable of breaking into smooth, even blocks.[1]

The spring and the remnants of that camp are surrounded by a chaos of great broken sandstone boulders, arranged like a mythical giant's fallen house of cards, with the smooth surfaces covered by art. As many as 2,500 separate etchings on some 200 sandstone tablets depict a swirling congregation of antelopes, elephants, leopards, lions, ostriches, rhinos, warthogs, zebras—and giraffes—as well as some humans, the occasional animal and human hand or foot print, and a number of purely abstract forms and designs. The representations are convincing and accurate and yet boldly stylized. There are rhinos, for example, with impossibly long upturned horns, tapered and fragile. There is a lion with a preternaturally long tail that curls back and then up and finally terminates in a leonine paw print. A giraffe stands on finely tapered footless legs that look like wisps of smoke rising from a fire. Another giraffe, elsewhere in the stone, stands proudly with a five-pointed head, five projections (two ears and two horns on top, a smaller horn pointing back) that strangely evoke the five digits of an outstretched human hand.

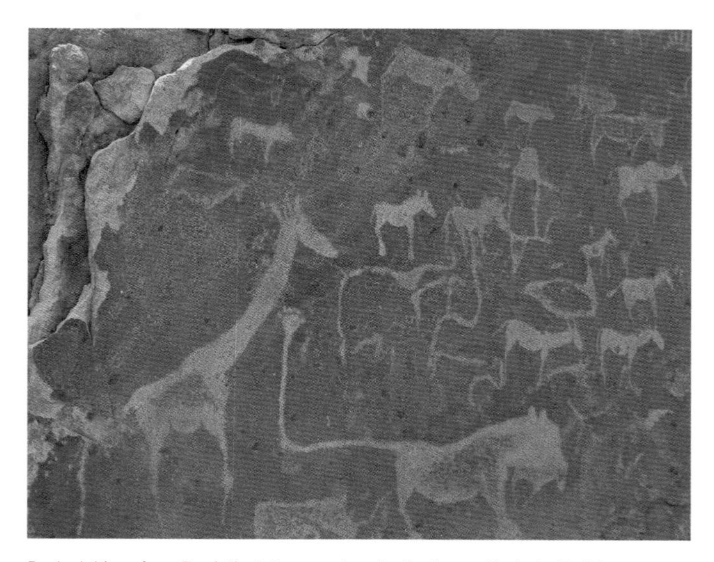

Rock etchings from Twyfelfontein camp done by Bushmen. Photo by Karl Ammann.

Before writing came art, and so it is art that draws us back to the beginning of memory. Africa is covered with such memory, which has been painted on or carved and chipped into rock. The art embraces the artists themselves and their people, and it embraces the animals people lived with, the animals they saw and dreamed about and hunted when hunger so required.

The art can be found far to the north, from the western edges of the Nile River all the way west across the Sahara, from there down to the eastern middle of Africa, and down again to the south. The northern art reminds us that the Sahara Desert was once, before a shift in climate that happened four to six thousand years ago, wetter and richer and far more hospitable to large mammals and large-mammal hunters than it is today. Giraffes are depicted there, often, in the context of hunting and trapping. But the southern carvings and paintings, all done by Bushman artists and revealed in thousands of different sites across Africa's great southern foot, evoke, I think, a more ancient life that took place under the sun and stars within a coherent and whispering cosmos.[2]

The Bushmen were despised by the first white settlers in Africa, who saw them as wild men with clouded minds and filthy ways, a people inherently incapable of grasping the higher logic of Christian and colonial authority, with (in the words of one early missionary) "a soul debased, it is true, and completely bound down and clogged by his animal nature."[3] They were "savages," to repeat the calumny used by Sir John Barrow in his memoir of explorations in southern Africa done more than two hundred years ago. Barrow, though, was expressing a common prejudice, and he probably did so ironically, while describing his early discovery of the glorious art surrounding a Bushman camp, art so forceful and spirited, so accurate and yet expressive, that, he wrote with a critic's understated certitude, "worse drawings . . . have passed through the [European] engraver's hands."[4]

Barrow recognized the skill and intelligence involved in such art, and he responded to it in aesthetic terms. This art is not the fading remnant of a feeble attempt at decoration or of casual vandalism, the graffiti of bored teenagers. It is the studied production of an active mind. Barrow saw beauty, and he recognized training and skill. That is an appropriate response, yet it is inappropriate to imagine that the Bushman artists intended

> The fires were scarcely extinguished, and the grass on which they slept was not yet withered. On the smooth sides of the cavern were drawings of several animals that had been made from time to time by these savages. Many of them were caricatures; but others were too well executed not to arrest attention. The different antelopes that were there delineated had each their character so well discriminated, that the originals, from whence the representations had been taken, could, without any difficulty, be ascertained. Among the numerous animals that were drawn, was the figure of a zebra remarkably well done; all the marks and characters of this animal were accurately represented, and the proportions were seemingly correct. The force and spirit of drawings, given to them by bold touches judiciously applied, and by the effect of light and shadow, could not be expected from savages; but for accuracy of outline and correctness of the different parts, worse drawings than that of the zebra have passed through the engraver's hands. **—SIR JOHN BARROW, 1806**

these works to be, in the European way, aesthetic productions that might be bought or sold or traded, thereby distinguishing the artist as an individual. Nor is there any clear suggestion in this art of the simplistic tit-for-tat of sympathetic magic: the effort to capture or freeze game animals symbolically with the fervent belief that an artist's triumph can become the hunter's.

Yes, individual artists must have been particularly skilled, and surely this art would have generated aesthetic pleasure as well as a sense of wonder or magic. But its primary purpose may have been collective rather than individual, and it must have worked in the same way that stained-glass windows did for illiterate medieval Christians: as a cultural expression, a shimmering communal statement in which the ways and logic of a people within their cosmos were confidently remembered, rehearsed, and realized.[5]

Our guide at Twyfelfontein, a slender and composed young Damara woman who introduced herself as Thekla Tsaraes, explained that the carved rock art was done by Bushman shamans who had gone into a trance. During the trance, she said, they used their art, those ethereal representations of animals, as a route of entry into the spirit world. The giraffes, for instance, were usually shown without their hooves, with their legs drawn away into long, thin lines expressing the shaman's experience of rising in the air when he enters a trance. Sometimes a giraffe etching would be twisted, in the way a shaman feels his own body changing, transforming as he enters the spirit world.

When she spoke of the Bushmen, Tsaraes said "Boesman," and her English was sometimes hard to follow. "So the Boesman people," she said, "have used their footprints to enter the solid rock without being seen." When I pressed her about the giraffe images, she declared, "Sometimes even the giraffe is regarded as a holy animal. They believe it's close to the clouds and is bringing down the rain." And when I asked her how we could know such things about people who lived so long ago, she responded that anthropologists had studied their culture.

It is true. We know a good deal about the cultures of surviving Bushman groups from the work of twentieth-century anthropologists. None of those survivors made the art, however, and the primary source of knowledge about the art-making Bushmen comes from the nineteenth-century labors of Wilhelm Bleek, a German linguist living in South Africa. Bleek was interested in studying the several languages of Africa's First People, and when, in 1870, he learned that some /Xam Bushmen were imprisoned in Cape Town for various petty crimes, he convinced the colonial governor to release a number of them to his care. One of them, a man named //Kabbo who was, in Bleek's assessment, "a gentle old soul, lost in a dream-life of his own," proved to be his most prolific informant, although the other /Xam also contributed.[6] They lived in Bleek's house, taught him their language, and in the process described their lives and vanishing culture.

The /Xam lived in extended family groups of perhaps a half dozen to two dozen people, who would temporarily settle near a spring or water hole. They built their small huts far enough from the water to avoid frightening the animals, who also congregated around water, and they relied on a second spring or water hole for the change of seasons and the inevitable drying-up of the first. Getting to the second might require two or a few days' trek across arid lands, with the migrating group carrying water inside ostrich eggshells.

They were hunters and gatherers, with the women gathering vegetable foods and the men hunting for meat using small bows and light, poison-tipped arrows. The /Xam poisons were lethal but very slow acting, which meant that the hunter had to track his wounded quarry for hours or even days.[7] Tracking, then, was an essential skill for these hunters and is expressed in the animal-track motif of so much of their art.

But the /Xam worked to control their fickle and often hostile environment through shamanism, which is even more of an essential theme for the art. All-night dances brought some of the men, carrying sticks and wearing rattles made of dried seed pods or pebble-filled springbok ears, into a trance state. The dancers, trembling, sweating, bleeding from their noses, became charged with a potent energy that seemed to boil out from within. Through succumbing to this energy they experienced their own death, leaving their physical bodies in order to manipulate the occult forces of the world beyond. They became shamans, in other words, and they used their newly acquired powers to work on three interconnected problems having to do with health, game, and rain. Shamans who acquired the power of healing might pull the illness out of a stricken person and into themselves, then sneeze it out along with a bloody discharge, which was then wiped onto the ill person with the theory that its smell protected against evil. Game shamans—the rock art sometimes shows them wearing caps made from the scalp of an antelope, the ears sewn to stand upright—worked to control the movements of antelope herds and confuse the trickster deity, /Kaggen, who liked to protect the special animals. And finally, the rain shamans tried to outsmart and catch certain mythical rain animals, whose blood or milk, when spilled, would be transformed into water that fell as rain.

That, in any event, is what I learned at Twyfelfontein and later from considering a handful of books on the subject. I also spoke about such things with Elizabeth Marshall Thomas, author of the anthropological classic *The Harmless People* (1958) and, more recently, *The Old Way: A Story of the First People* (2006), both of which draw on her experiences as a girl visiting and living among four language groups of the still surviving Kalahari Desert Bushmen.[8] She knew nothing about the rock art, Thomas told me, since the Kalahari Bushmen did not do that kind of art. Their art was in their music—and, for the men, in their hunting and the mythlike stories they told about hunting.[9] Also, she added, none of the Bushman groups she knew had shamans, at least not in the sense of someone being an elite, professionalized healer.

It was true that certain men, sorcerers by reputation, were said to possess the power to fly through the air and enter the body of a lion. But any man could become a healer, and the powers for healing would emerge during their all-night dances. As with the /Xam, the Kalahari women sat in a circle and clapped, sharply and rhythmically, while the men danced in a circle around that circle, wearing rattles on their legs. At some time during this dance, often around dawn, a number of the men would acquire the power to draw away sickness. "Several men might fall into trance," Thomas said. "Then they'd come over and put one hand on your back and one on your front, and they'd suck away the something bad that was in you. They'd suck it into themselves, and then they'd scream it into the air. And what they sucked out they called 'star sickness.' They did this for real illnesses too, but a star-sickness healing would be for things that cause jealousy, ill-will: bad things that can make a group disintegrate."

The Kalahari Bushmen's notoriously fickle gods occupied the horizon, especially during those shifting, magical moments just before sunrise and sunset, and the spirits of the dead served those gods. The spirits were anyone who had died, and they were all around. They lived in this world, invisible yet leaving fine trails in the air, like spiderwebs floating about fifteen feet off the ground. People talked to them, sometimes cursed at them, and you knew where the spirits were because you could see where people focused their eyes and projected their voices.

Did animals become spirits? I asked. Thomas said she never heard that they did, but it was true that men who had gone into a trance would swear at the spirits of the dead and also at the lions—but the lions were really there, in this world. They weren't spirits. They were just animals who shared a water hole with the people.

Still, giraffes and certain antelopes ("the large ones, the ones that are a hunter's special prize") shared with people something called *n!ow*. This was a very mysterious thing that had to do with the weather. A man could urinate into a fire, and his *n!ow* would interact with the fire, causing a change in the weather. When a woman gave birth and her amniotic fluid hit the ground, there would be a change in the weather. Similar things happened when a hunter spilled the blood and thus the *n!ow* of a giraffe or a large antelope onto the ground. This remarkable substance or energy or force possessed by humans and giraffes and the five next-biggest game animals (elands, wildebeests, kudus, gemsboks, and hartebeests), this *n!ow*, could change the weather.[10] "It's so remote from anything we have anything to do with that it's hard to understand," Thomas concluded, "but the important thing to me is that people had it, and the big antelopes they hunted had it, and the giraffes. Period. No one else had it. It was an important characteristic."

The photograph introducing this chapter was taken in the situation described in the opening text: a moment just before dawn when Karl and I came across three giraffes lying down in the Masai Mara. The photo includes two of those three.

The images below (all of Masai giraffes in Kenya's Masai Mara) present an imagined sequence following that predawn wake-up: awakening into a full morning, a midday storm and heavy rain, and a lifting of the weather into late afternoon and evening. The overarching theme is giraffe ethereality as expressed through faint or hidden images and long-distance silhouettes. Note that the giraffes standing in the rain, at first rendered ghostly by the heavy downpour, are driven by the angle of the wind and rain to orient themselves in the same direction. Two of the later images include the forms of hidden giraffes.

CHIMERAS

ONE WINTER DAY in the third decade of the third century BC, Ptolemy II Philadelphus, the Greek king of Egypt, presented in the capital city of Alexandria the biggest parade in history.[1] This Grand Procession, as it has been called—an all-day event that began with the morning star and ended with the evening star—was intended to express to the entire world the rising power, piety, and glory of old Egypt under her new Greek ruler.

To project power was essential given that the enemies of Egypt were themselves powerful and menacing. The military message of the Grand Procession would have been impossible to miss, since its final portions included a fully armed and armored display of close to 48,000 foot soldiers and more than 23,000 cavalry mounted on prancing steeds. A middle portion of the parade displayed ninety-six elephants marching four abreast and harnessed, in that formation, to twenty-four enormous chariots. These would have been just a portion of Philadelphus's war elephants, barged downriver from their training quarters in Memphis for the parade. Armored with quilt-wrapped metal or fire-hardened leather, armed with sharp metal blades affixed to their tusks, taking onto their backs skilled javelin throwers, and emotionally fortified at the last minute with buckets of red wine, the parading pachyderms could readily be transformed into fearsome battle tanks supporting phalanxes of Egyptian infantry and cavalry.

Yes, it was a military parade. But the Grand Procession was also a religious one, an unfolding projection of pious obeisance to the deeper powers of the universe—powers, of course, as the Greeks understood them, and in this instance concentrating on the grape-growing, wine-making, mind-expanding powers associated with the god Dionysus, from whom the Ptolemies had begun claiming ancestral descent.

In honor of Dionysus, the parade was opened by actors dressed as the god's companions and allies, the older Sileni (who restrained the crowds) and the youthful Satyrs (wearing artificial tails and outrageously large phalluses and carrying torches of gilded ivy). Next came incense—aromatic clouds of frankincense, myrrh, and saffron wafted into the air—followed by more costumed actors representing further aspects of the Dionysian myth. Then came a cart more than twenty feet long and a dozen wide, drawn by one hundred eighty men and transporting a fourteen-foot-high statue of Dionysus himself, dressed in a purple robe with a draped fold that was purple interwoven with threads of gold. Shaded by a fruit-laden, vine-laced canopy and standing before a golden table holding numerous gold vessels, the statue poured a continuous libation from an enormous, two-handled gold cup. Other celebrations to grapes and wine and the wine god followed, including a cart drawn by three hundred men transporting a giant wine press filled to the brim with ripe grapes. Sixty Satyrs sang stirring songs while stepping squishingly onto the grapes. Now came six hundred men pulling a bigger cart containing a double-spouted wine vessel made of leopard skins and pumping seasoned wine into the street as it passed. One hundred twenty Satyrs and Sileni followed, carrying gleaming vessels of solid gold, and they were followed in turn by a giant silver bowl on a cart drawn by six hundred men. More statues passed by, including an eighteen-foot Dionysus lying on an elephant, a golden phallus one hundred eighty feet long, a gilded thunderbolt sixty feet long, and a gilded shrine sixty feet in circumference. Thus passed the iconography representing the spiritual moorings of Egypt's latest rulers.

The glory of Philadelphus's Grand Procession was largely an exuberant expression of economic triumph, a day-long presentation of one gleaming treasure after the other, demonstrating—with a veritable river of jewels and silver but above all gold, gold, and gold—the astonishing wealth that the Greek king had accumulated on behalf of himself and his subjects.

But Philadelphus's glory was also expressed more diversely in the form of a living zoological wealth, as he revealed his command over the many creatures, both dangerous and docile, he had gathered from all over the known world. Thus, the ninety-six elephants lined up in rows of four and harnessed to chariots were followed by a parade of two-animal chariots drawn by one hundred twenty goats, thirty hartebeests, twenty-four saiga antelopes, sixteen ostriches, fourteen oryxes, fourteen Persian wild asses, eight Asian wild asses, and eight horses. Small boys and girls rode in the chariots, the boys dressed up as charioteers, the girls as warriors. The procession also included twenty-four hundred dogs of various breeds and—destined for ritual slaughter—four hundred fifty sheep and forty-six head of cattle. Meanwhile, men carried trees laden with birds and animals of many sorts. More men carried cages containing guinea fowl, parrots, peacocks, pheasants, and other exotic birds. And yet more paraders guided and restrained nineteen cheetahs (three of them cubs), fourteen leopards, four lynxes, one bulky bear, one hulking rhinoceros, and one no doubt very elegant giraffe.[2]

—

The giraffe may have been Ptolemy Philadelphus's most singular surprise. That animal was, as far as the historical record reveals, the first giraffe to reach Egypt for nearly a thousand years—with the closest known predecessor harking back to the time of Ramses the Great, who, according to a painting on a temple wall, received a giraffe as tribute from Egypt's southern neighbors along the Nile during his reign as pharaoh from 1279 to 1213 BC.[3]

A thousand years may seem like a long time—approximately the remove between ourselves and such hoary historical events as, say, the Norman Conquest of England. Why so long? And what were the consequences of such a distance in time?

The first question has a simple and obvious answer. It took so long because it was so difficult. Giraffe ancestors originally came from Asia, and they moved into Africa during a migration that happened around seven million years ago. This event could have occurred gradually, perhaps over hundreds or thousands of years. From a forgotten, mysterious place far to the north and east, so we can imagine, they migrated, moving, always moving, through life and birth and death, following not a vision but the scent of steady moisture and the food it would foretell. They ambled across the Arabian land bridge and on to the very edge of Africa where, through an open door, there appeared the promise of sufficiency.

Through the open door they passed. But then, over time, the door began to swing shut behind them. The rains, in other words, began to fail. The monsoons that had once pulled endless clouds of water away from the seas, carried them overland and into the northern reaches of a continent, lost their energy. The weather and then the climate changed. The welcoming wet lands of northern Africa, that great green cap of a continent, turned drier. Rivers became seasonal, then unreliable, then nonexistent, leaving a faint feathery trace across the sandblasted land. Vegetation turned brown, withered, became sparse. It died or survived by adapting to this far less merciful world. The earth turned barren and hostile, becoming a seething wilderness of sand piled against rock, with a ragged palisade to the north and west. In the west, the wilderness broke at last into cliffs that slipped into the boiling ocean. To the east, it descended finally into the shallows of a warm salt sea. The great green cap of Africa became a great brown cap. At 3.32 million square miles, this impossible wilderness now approaches the size of the United States. It is what you and I, harkening back to the Arabic word for *desert,* call the Sahara.[4]

So the Sahara was once wet and green. In fact, it was several times wet and green, that condition occurring in a slow and steady oscillation with dry and brown, with the latest wet period receding some six to eight thousand years ago.[5] That is recent enough that we can examine the cultural evidence of teeming wildlife in a vast plateau of southeastern Algeria, the Tassili n'Ajjer. Here, in one of today's most unforgiving environments, we can contemplate a gallery of some 15,000 rock paintings and etchings, many of them drawing us back to a time when the place was a grassy savanna washed by rivers and streams, when such water-dependent species as crocodiles and hippos were abundant, when the many game species included large mammals such as giraffids and giraffes.

Giraffes are capable of surviving in very arid environments. Karl and I sighted them foraging along dried-up riverbeds and running past giant sand dunes in the deserts of northwestern Namibia. They can endure for a long time, perhaps indefinitely, without actually drinking water, as long as they find food that includes enough moisture. In that sense, they resemble dromedary camels: famously tolerant of extreme aridity and used for transportation in the Sahara. Both giraffes and camels have slit-like nostrils, which may be an evolutionary adaptation to windblown sand. And again like camels (and only a few other large-bodied, warm-blooded species), giraffes have a thermoregulatory system that allows their body temperature to drift. Giraffes thus have less need to expend energy keeping themselves cool on exceptionally hot days and warm on very cold nights.

Nevertheless, as that latest shift in climate began transforming the Sahara from grassy savanna to barren desert, the giraffes and a number of other large mammals living in the Sahara went extinct or began retreating south. The door, as I say, was closing.

The final crack of that closing door was the Nile Valley. Giraffes lived along the lower Nile as late as 3800 to 3400 BC, when early Egyptians were producing pottery and carving ivory knife handles that sometimes depicted them. In the fifth dynasty (2750 to 2625 BC), a hunting scene that included a giraffe was carved in bas relief on the tomb of King Unas at Sakkara. The tomb of Ukt-Hop in the twelfth dynasty (2000 to 1780 BC) portrayed, alongside the hunting dog and the arrow-pierced antelope, another hunted giraffe. But that was the last. Giraffes may have gone locally extinct in Egypt around four thousand years ago.[6]

Giraffes were still alive, of course, enduring in richer and less settled lands to the south. But contact between Egyptians and their southern neighbors along the Nile Valley was limited by the difficulty of passing through the river's cataracts and the forbidding deserts and mountains surrounding them.

Queen Hatshepsut, the fifth pharaoh of the eighteenth dynasty (1508 to 1458 BC), overcame that limitation by organizing trading voyages on the Red Sea. Hatshepsut sent five large ships, each around seventy feet long and crewed by more than two hundred men, down the Red Sea to what was called the Land of Punt at its southern end, thereby enriching the Egyptian treasury with such coveted items as gold, ivory, ebony, and myrrh. The ivory came from elephants, but Egyptian records of the expedition indicate that Punt was also a land of baboons, hippopotami, and leopards, while carvings on Hatshepsut's tomb indicate a live giraffe brought back from Punt as a gift or tribute to the queen.[7] A few other pharaohs received giraffes as tribute from lands along the southern Nile, including King Tutankhamun and, as I mentioned earlier, Ramses the Great. Then came the thousand-year gap.

In Egypt, giraffes may have simply ceased to exist, even as lonely captive animals kept alive in some pharaoh's garden, after the death of Ramses. How much longer would it take before that absence of physical fact would become an absence in memory? And as Egypt was drawn, during the next millennium, into the orbit of a larger Mediterranean culture, how would that amnesia come to affect the Western world of classical times?

The Greek historian Herodotus (484–425 BC) traveled extensively in the Mediterranean and the Middle East, and he spent time in Egypt around 454 BC. Herodotus regularly included in his nine-volume *History* the often strange and sometimes fanciful accounts of travelers to and from exotic places. Herodotus had nothing to say about giraffes.

A century later, Aristotle (384–322 BC) wrote with astonishing catholicity and an often impressive precision about zoology and natural history, producing the world's first zoological encyclopedia. Firsthand knowledge about elephants arrived in the Mediterranean basin during Aristotle's lifetime, brought from the east by his erstwhile student, Alexander the Great. Aristotle's extended and detailed commentary on elephants was generally accurate enough to remain relevant until well into modern times. He, too, failed even to mention giraffes.

By the time of Herodotus and Aristotle, giraffes may simply have ceased to exist for the Mediterranean world, physically or conceptually. Neither whispered nor speculated about. Not remembered or dreamed about. Not even imagined as, say, a fantasy creature of many parts, a strange and mythical chimera rising out of some wavering obscurity at the far end of the Nile.

—

Ptolemy II Philadelphus reopened the door to giraffes not so much out of curiosity or a sense of adventure, although such inclinations surely were part of the equation. He did it, perhaps primarily, out of an ordinary, old-fashioned fear: the reasonable fear of being annihilated in war.

Elephants had come to the Mediterranean in 325 BC, when Alexander the Great returned from India trailing, as tribute extorted in peace and booty seized in war, around two hundred fearsome war elephants. Establishing his imperial capital at Babylon, the young man settled down to rule his empire from a golden throne in a tented pavilion, surrounded by a bodyguard of Persian and Greek soldiers and a central corps of war elephants. He appointed an *elephantarch,* a commander of the elephants, and he moved to integrate the animals more fully into his own forces. Elephants would be an essential part of his new war machine and would lead him to future conquests and even greater glories . . . or so the young man may have dreamed. Unhappily, he died of a sudden illness at the age of thirty-two, before any such dreams could be realized.

Alexander's death was also the death of an empire. In a series of campaigns waged south into Palestine and Egypt, west into Turkey and Greece and Italy, his several former generals and viceroys fought each other ferociously, using the latest and most impressive weapon anyone from the West had ever seen: elephants, the battle tanks of the classical world.[8]

Among the competing inheritors of the shattered empire were Seleucus and Ptolemy I. Seleucus established his military headquarters in northern Syria, where he stabled his own trained corps of about five hundred elephants recently acquired from India. A glorious image of elephants pulling Seleucus on a chariot was stamped onto a coin of the realm, while the real Seleucus and his real elephants in Syria must have disturbed the sleep of Ptolemy I in Egypt, who could marshal only a few dozen of the creatures with which to defend himself and his piece of empire.

Ptolemy I acquired some of the imperial charisma by stealing Alexander the Great's body as it was being shipped back to Greece, then installing the rotting corpse in a gold sarcophagus inside a grand mausoleum at Alexandria, the newly founded city at the Mediterranean tip of Egypt. But charisma was no substitute for elephants, and, for Ptolemy, the big problem with elephants was how difficult they were to acquire. Indeed, the fact that all war elephants came from India and were members of the Asian species now put Ptolemy and his pathetic pack of pachyderms at a terrible disadvantage, since Seleucus controlled the route east to India.

Ptolemy's solution was to look south, to the shady glens and shadowed forests of deeper Africa, Africa south of the Sahara, and to acquire African elephants. After Ptolemy's death, his son Ptolemy II Philadelphus took up the cause. It was already known that such animals could be found to the south of Egypt, but Philadelphus sent military and diplomatic expeditions down the Red Sea and up the Nile to discover more precisely where they were and how to get some. Some of those expeditions returned with reports on the geography, people, resources, and mercantile opportunities in lands south along the Red Sea, while others reinforced the long-standing ties between Egypt and the people living along the southern Nile, particularly in the kingdom of Meroë, in today's Sudan.

To Meroë, Philadelphus sent teams of Egyptian soldiers and Indian elephant trainers (mahouts) who, working with the local experts, began to capture, tame, and train elephants.[9] Meroë maintained its regular trade with Egypt, going north by way of the Nile, but the river's cataracts meant that the captured elephants could not be sent directly north on river boats. Instead, once they had been trained well enough to walk under the command of the Indian mahouts, they were marched east for several days until they reached the Red Sea, loaded onto specially designed (sturdy, flat-bottomed, shallow-drafted) elephant boats, then sailed north for several days to a few weeks, then marched west across the desert for about twelve days until they reached the Nile in Egypt. At the great river, the elephants were walked onto barges and floated downstream to Memphis, where they would be trained to serve in war.

Had he been a warrior like his father, Philadelphus might have been satisfied with the growing size and power of his new elephant corps in Memphis. But Philadelphus, for all his emotional and physical weaknesses, was imaginatively ambitious in ways his father was not. In the words of the first-century Greek geographer Strabo of Ephesus, the young Ptolemy was "of an inquiring disposition, and on account of the infirmity of his body was always searching for novel pastimes and enjoyments."[10]

The father had established the great museum and library at Alexandria, but the son brought in scholars and expanded those institutions until Alexandria was the center of learning for the Western world. Likewise, Philadelphus was not content with the acquisition of elephants, or even with the southern expeditions' secondary effect of opening new gold mines and expanding Egypt's trade and influence into Arabia, India, and sub-Saharan Africa.[11] Philadelphus encouraged the capture and transport of all sorts of exotic animals, bringing them back as specimens for his growing menagerie in Alexandria.[12] None of these animals would be useful in the way he expected elephants to be, of course, but for Philadelphus they were parts of a living zoology and at the same time impressive collector's items that would contribute to the charisma any great ruler strives to maintain.

The Greek historian Diodorus wrote that Philadelphus was "interested in capturing elephants" and "gave liberal rewards to those who engaged in the strange hunts for these powerful animals. He spent large sums of money on this hobby, and collected a considerable number of war elephants; moreover he acquainted the Greek world with other strange and unheard of animals."[13] Philadelphus's giraffe, exhibited before the world that winter's day in the third decade of the third century before Christ, was one of those "strange and unheard of animals."

—

Giraffes, in truth, were so strange and unheard of that neither the Greeks nor the Egyptians knew what to call them, and so the Greeks were forced to invent a name.

The difficulty of choosing a name for this creature was like that of naming anything that appears strikingly outside the usual categories—like, for example, naming an unusual sound or a peculiar smell. Without the guidance of comparative examples or rational categories, one resorts to creative metaphor.

The namers of Philadelphus's giraffe could have been the leaders of an early capture expedition to the south. Perhaps they were Greek translators chatting casually with Egyptian crew members on an elephant boat transporting the just-captured animal north on the Red Sea. Whoever they were, the namers would have had the same problem—what do you call something that defies the known categories?—and so they named him using the most telling associations they could think of. The creature had a rather camel-like face, and he seemed tall and lanky like a camel. At the same time he had those peculiar spots. Not at all like a camel. More like a leopard.

They called this new animal a *camel-leopard*—or, as the English translators more often represent it, a *camelopard*. And since the Greeks, like the rest of us, had trouble distinguishing a figure of speech from a figure of fact, they came to imagine that camelopards were the natural product of a camel mating with a leopard. They were hybrids in name and fact, fantastic chimeras taken from the depths of sub-Saharan Africa.

The Greek historian and geographer Agatharchides of Cnidus, writing around 104 BC, refers to giraffes in his natural history of human tribes and exotic animals living in the harsh regions west of the Red Sea. His original text was lost, but not before later authors extracted passages, such as the reference to giraffes as animals "which the Greeks call *camelopardalis*, a composite name which describes the double nature of this quadruped. It has the varied coat of a leopard, the shape of a camel and is of a size beyond measure. Its neck is long enough for it to browse in the tops of trees."[14]

Strabo of Ephesus, however, writing in the next century (and citing the work of a geographer named Artemidorus, whose work has been lost) insisted that "camelopards . . . are in no respect like leopards":

> for the dappled marking of their skin is more like that of a fawnskin, which latter is flecked with spots, and their hinder parts are so much lower than their front parts that they appear to be seated on their tail parts, which have the height of an ox, although their forelegs are no shorter than those of camels; and their necks rise high and straight up, their heads reaching much higher than those of camels. On account of this lack of symmetry the speed of the animal cannot, I think, be so great as stated by Artemidorus, who says that its speed is not to be surpassed. Furthermore, it is not a wild beast but rather a domesticated animal, for it shows no signs of wildness.[15]

Strabo was wrong, of course, in insisting that giraffes are domestic animals. But he was right in recognizing their gentleness. And his second error, that they are not especially fast runners, confirms the already obvious fact that he never saw a giraffe running free. Still, the overall precision and self-assurance of Strabo's description do suggest that he had either seen a live giraffe—albeit one in captivity—or spoken at length with someone who had.

If so, which giraffe might that be?

Strabo traveled widely in the eastern Mediterranean; he came to Rome around 44 BC, and he took part in a Roman expedition up the Nile into southern Egypt in 25–24 BC—just a few years after Cleopatra, the last of the Ptolemies, committed suicide, thereby ending the rule of the Greeks. Egypt became a Roman colony. But Cleopatra, during a happier time and following the fashion that originated with Philadelphus two centuries earlier, probably kept exotic animals on the palace grounds in Alexandria. Perhaps Strabo saw one of Cleopatra's giraffes, after she was gone and Romans were occupying the palace. . . . Or perhaps Strabo was in some way familiar with the live giraffe Julius Caesar displayed in Rome as part of his 46 BC triumph, which happened while Cleopatra was still alive: Caesar's guest in Rome while still Egypt's queen.

Romans had by then become accustomed to seeking live animals for their increasingly grand and fantastically bloody animal spectacles. Exotic live animals thus became part of the normal economic exchange between Rome and her colonial empire. From the colonial wildernesses, then, specimens were routinely captured, caged or otherwise restrained, placed on any available boats, and shipped to Rome.[16] Giraffes, though, only came into the Mediterranean from Egypt, taken as ever from far to the south—Ethiopia, for instance—and passed as usual down the Nile, portaged across the cataracts, brought into the country as a prize for the royal collection in Alexandria. To be sure, Caesar's soldiers could have acquired one directly, from a commercial or diplomatic exchange with people to the south of Egypt. But I agree with biographer Stacy Schiff, author of *Cleopatra: A Life* (2011), who argues that the Egyptian queen herself was probably the original owner of Caesar's giraffe.[17]

Cleopatra left for Rome in the summer of 46 BC, transported, along with her and Caesar's one-year-old son, Ptolemy Caesar, plus essential servants, in a naval galley: likely a swift-running 120-foot trireme powered by square-rigged sails and 170 oarsmen. The royal boat proceeded out of the Alexandrian harbor accompanied by a grand flotilla of supporting vessels, enough to transport the royal retinue and bodyguard and a large personal and institutional staff—astrologers, priests, philosophers,

His size was about that of a camel; his skin, like that of a leopard, was decorated with spots in a floral pattern. His hindquarters and belly were low and like a lion's; the shoulders, forefeet and chest were of a height out of all proportion to the other members. The neck was slender, and tapered from the large body to a swanlike throat. The head was shaped like a camel's and was almost twice as large as that of a Libyan ostrich. The eyes were brightly outlined and rolled terribly. His heaving walk was unlike the pace of any land or sea animal. He did not move his legs alternatively, one after the other, but first put forward his two right legs by themselves, and then the two left, as if they were yoked together. Thus first one side of the animal was raised, and then the other. Yet so docile was his movement and so gentle his disposition that the keeper could lead him by a light cord looped around his neck, and he obeyed the keeper's guidance as if the cord were an irresistible chain. The appearance of this creature astonished the entire multitude, and extemporizing a name for it from the dominant traits of his body they called it camelopard. **—HELIODORUS, CA. AD 220**

advisors, physicians, secretaries, cooks, and so on—as well as loads of personal effects and opulent gifts of the sort one would expect from the world's wealthiest person, the great monarch of a great civilization, an official goddess, and Julius Caesar's lover. Those royal gifts could very well have included a giraffe.

Cleopatra should have been satisfactorily ensconced in Caesar's country estate, just outside the city walls, by the time the Roman dictator opened his eleven days of triumph, on September 21. The festivities consisted of grand parades, enormous feasts, spectacular entertainments, bloody gladiatorial contests, horse races, forty elephants lighting up the night with forty flaming torches held in their trunk tips, lions by the hundreds, leopards, panthers, baboons, monkeys, flamingoes, ostriches, parrots . . . and one giraffe.

The clearest report we have of that tall and undulatory beauty, the first of his or her kind ever to set foot on the European continent, comes from the poet Horace (Quintus Horatius Flaccus, 65–8 BC), who chided his fellow Romans as "a throng gazing with open mouth" taking foolish pleasure in the spectacle, particularly Caesar's giraffe, which was "a beast half camel, half panther."[18] Caesar climaxed that memorable presentation, unfortunately, with blood: sacrificing the giraffe to hungry lions in an arena.

Later Roman worthies would parade a few more giraffes before the plebeian masses—ten of them together in the circus of AD 247, when the emperor Gordianus III celebrated Rome's first thousand years[19]—but the essential rarity of giraffes may well account for the paucity of accurate descriptions we have. What notably remains, in the classical record, is a pair of evocative passages from two writers of the early third century AD. The first writer, Oppian of Apamea (in Syria), portrays giraffes, in his poem on hunting, *Cynegetica,* as animals of "a hybrid nature and mingled of two stocks," the camel and leopard. The poem is dated by its dedication to the Roman Emperor Caracalla, meaning it would have been finished a few years after AD 210.

The second work is the *Ethiopian Romance,* a fictional entertainment that appeared around AD 220 and was written by someone using the pseudonym Heliodorus. Set in a North African world as imagined to have existed several centuries earlier, a giraffe appears in a grand procession marking the conclusion of a major war. Here, the triumphant Ethiopian king Hydaspes receives tribute from defeated enemies as well as congratulatory gifts from his friends and allies, the latter including the Auxomites, who offered "a marvelous animal of extraordinary appearance."[20]

"Chi-me-ra: 1) In Greek mythology, a fire-breathing female monster with a lion's head, a goat's body, and a serpent's tail. 2) Any mythical animal with parts taken from various animals" (*New Oxford American Dictionary*).

The ancient Greeks may have originally decided to call giraffes camel-leopards as a quick and simple reference to physical appearance: an animal with camel-like face, camel-like gait, camel-like legs, who is, however, covered with spots roughly suggesting a leopard. The name could have been nothing more than an easy shorthand for appearance. The description was sooner or later accompanied by a theory of giraffes as true hybrids: a remarkable cross leading to the rather miraculous convergence of physical features in the way that chimeras were imagined as miraculous conglomerations.

The opening photograph for this chapter shows a lone Masai giraffe standing near his reflection in a pool of water. The sequence below is a visual fantasy based on a concept of chimeras and the theme of reflections and resolutions, separations and convergences. The photographs show both Masai giraffes (in Masai Mara, Kenya) and reticulated giraffes (in the Samburu National Reserve, Kenya). Note how different the patterning is between the Masai and the reticulated: two groups of giraffes who live in the same general area of East Africa but have not interbred for more than a million years.

UNICORNS

IN THE HISTORICAL ANNALS of the Chinese, the earliest known reference to Africa appears in the *Yu-yang-tsa-tsu,* written by the scholar Tuan Ch'eng-shih, who died in AD 863.

Relaying stories and information that had been provided by travelers from the West, Tuan described a land called Po-pa-li, which probably corresponds to a coastal portion of today's northern Somalia. This hostile, faraway land was home to some strange animals, the scholar wrote, including "the camel-crane" (ostrich), the "mule with red, black, and white stripes wound as girdles around the body" (zebra), and "the so-called *tsu-la,* striped like a camel and in size like an ox. It is yellow in colour. Its front legs are five feet high and its hind legs are only three feet. Its head is high up and is turned upwards. Its skin is an inch thick." Both these odd quadrupeds—the striped mule and the tsu-la—are "variations of the camel," which "the inhabitants are fond of hunting and from time to time they catch them with poisoned arrows."[1]

The tsu-la, then, was probably a giraffe.

At various times following that earliest reference, the Chinese traded with African countries through intermediaries, particularly as, during the tenth and eleventh centuries, their trading ships sailed as far as southeastern India to exchange their own valuables for such luxury goods as elephant ivory, rhinoceros horn, pearls, and precious aromatics. But the Chinese would not see or touch an actual giraffe until AD 1414.

That gorgeous creature arrived during China's Age of Exploration, a great if brief period that lasted from AD 1405 to 1433 and was inspired—or commanded—by the Ming emperor Yongle, who opened China to an assertive form of maritime trade with countries to the south and west. This new orientation may have been a natural consequence of the disintegration of the Mongolian Empire, which ended the Silk Road and an extensive overland trade between China and countries to the west. Under Yongle, the Chinese turned to the seas in seven enormous expeditions that eventually reached halfway across the world, through the Indonesian archipelago to India, the Arabian Middle East as far as Mecca, and on to the eastern shores of Africa as far south as the coast of today's Kenya.

Guided by magnetic compasses and complex star charts, the expeditions carried huge quantities of Chinese-made goods—copper and iron products, furniture and porcelain, cloth and silk and paper, sugar and salt—and returned to China with foreign envoys who oversaw the presentation of gaudy treasures to the emperor and his court while also profiting through commerce outside the court.

Nearly sixty years after the last of these fleets returned to port in Nanjing, Christopher Columbus sailed from Spain, leading a brave expedition that, in a disoriented search for the oriental Old World, would accidentally stumble onto an occidental New World. Columbus's three-ship expedition included a crew of about 120 men; his flagship, the four-masted *Santa Maria,* was approximately 80 feet long. By contrast, Emperor Yongle's first maritime expedition (commanded by Zheng He, a thirty-four-year-old Muslim eunuch who had previously served as Grand

Director of the Imperial Harem) carried a crew and army of around 28,000 men aboard 62 nine-masted ships and an additional 255 five-masted vessels. The larger ships, known as "treasure ships," were nearly 450 feet long and 190 feet wide.

So the full imperial fleet would have been a stirring or an alarming sight when it appeared on the horizon before the ports of various settlements and kingdoms in the Indonesian archipelago during the year 1405 and, from there, west as far as the southwestern coast of India. That first expedition returned in 1407 carrying several foreign emissaries along with their goods and treasures.

The fourth expedition left China in the fall of 1413 and sailed farther west than ever before, eventually reaching Hormuz, at the mouth of the Persian Gulf, and Aden, on the southern tip of the Arabian Peninsula at the mouth of the Red Sea. A subsidiary fleet from this expedition also sailed along the eastern coast of India to Bengal, where the sailors were greeted by a newly ascended Islamic king. It happened that envoys from the Islamic coastal settlement of Malindi, East Africa (in an area claimed by today's Kenya), were in the Bengal court at the time, having traveled there to offer their own tribute—some live giraffes—to the new king.

The Chinese visitors were clearly fascinated by those tribute giraffes. They encouraged the Bengali king to give them one, and they persuaded an envoy from Malindi to accompany that particular animal on one of their treasure ships that was returning to China before the rest of the fleet. China's first giraffe

Tribute Giraffe with Attendant. Chinese, Ming Dynasty, Yongle Period (1403–1424). Philadelphia Museum of Art.

thus weakly wobbled onto stable land on September 20, 1414. Within the year, at least one more giraffe was brought by sea from Milandi to Bengal and, from there, to the imperial capital at Nanjing, where it was presented to Emperor Yongle at some time before the full fleet returned home in August of 1415.[2]

The giraffes were introduced to the emperor and his imperial court as unicorns.

—

Ch'i-lin—the name given to the unicorn described in ancient Confucian texts—was what the mariners called the tall animals they brought back home, possibly because, one author speculates, they originally heard them described in the Somali language as *girin,* which may have sounded like ch'i-lin.[3]

It is also possible that Zheng He, the eunuch admiral of the fleet, firmly believed that the extraordinary animals he delivered to the emperor were actual ch'i-lin, actual unicorns as traditionally understood.

According to Confucian tradition, a ch'i-lin male, aside from his many other wondrous qualities, would be marked by a flesh-covered horn rising from the forehead. Giraffes have skin-covered horns, and some giraffe males develop a skin-covered median horn, a decisive knob or bump that appears at mid-forehead. Ch'i-lin could alternatively have two or three horns, as can giraffes. Confucian tradition also held that ch'ilin had a deer's body and cloven hooves, as well as the tail of an ox and, sometimes, the scales of a fish. A giraffe would probably pass that test as well, aside from the fish scales—or might a giraffe's markings actually resemble scales from a distance? Ch'i-lin were usually imagined to be white; but they could be gaily colored in red, yellow, blue, white, and black—not entirely unlike a giraffe. Ch'i-lin were associated with gentleness and goodness, qualities that would be at least superficially apparent in a giraffe; finally, ch'i-lin were revered as portents of good fortune brought about by a wise and benevolent ruler.[4]

The last imagined quality of a ch'i-lin suggests a third possible reason the giraffes brought back to China were presented as the miraculous unicorns of Confucian tradition: They could be used as propaganda bolstering the Yongle emperor's precarious claims to legitimacy.

Yongle was the Ming dynasty's third emperor. Or was he the second?

Yongle succeeded to the imperial throne in 1403 by raiding Nanjing at the head of an army of a few hundred thousand men, massing outside the city until one of his many brothers opened a city gate at night, whereupon his troops entered and overwhelmed the imperial defenses, setting fire to the palace and government buildings. The second Ming emperor, Jianwen, was (according to the claims circulated by Yongle) unfortunately and accidentally consumed in the flames.

Yongle moved to consolidate his own position as the new emperor, and his early acts included the usual—sorting friends from enemies, elevating the former, executing the latter—as well as arranging for an important recalibration of Ming history. This proved to be a major enterprise requiring that a select group of dedicated historians destroy all earlier records and accounts of the dynasty and then create a full replacement set of new records and accounts. Ultimately, Jianwen was expunged from the list of emperors altogether, which left the first and founding Ming emperor reigning for about four years past his own death. Yongle was then able to claim his honorable position as the second Ming emperor.

Jianwen had been the oldest surviving son of the founding emperor's first son. The founding emperor chose him based on the principle of primogeniture as described in the official *Ancestral Injunctions.* Yongle was merely a fourth son—and not even the child of his father's first consort, the empress Ma. Nevertheless, once he became emperor, Yongle made sure the official genealogy was rewritten, making him the son of Ma to fix the mother problem. And although he had not been the first of the founder's twenty-six sons, Yongle would argue that the Confucian principle of filial piety, along with the full legal code prescribed in the *Ancestral Injunctions,* allowed for a prince to intercede when an emperor was corrupted or overwhelmed by nefarious advisors. Yongle had been a prince. Yongle had interceded. Once he became emperor, his scholars would write the history that justified that intercession.[5]

In such a manner, Yongle became the most powerful man on earth: the civil, military, political, and spiritual head of an empire covering a stretch of real estate the size of Western

Europe and containing a growing population of perhaps 90 million people.[6] Yongle was among the most dynamic and influential rulers in Ming history, an era covering nearly three centuries. He was also a pretender to the imperial throne, an illegitimate usurper who would be concerned about the security of his position. Indeed, as the unhappy fate of his immediate predecessor starkly demonstrated, great power required great control. As the compelling example of his father more happily added, though, great control could sometimes be achieved through stimulating the emotions of hope, fear, and reverence.

Hope included the promise of advancement in the vast civil service bureaucracy and in the enormous military hierarchy. Fear was insured by an emperor's willingness to torture and execute anyone at court found wanting. Yongle's father had been responsible for the deaths of approximately 100,000 people who may not have seemed trustworthy enough. The founder also used his palace guard to create an infamous secret police with unchecked powers to arrest, torture, and execute.[7] Yongle was approximately as despotic as his father, and his palace guard was reinforced by a major spy network organized and conducted by the palace eunuchs, who by 1420 were organized into another extrajudicial secret police called the Eastern Depot.[8] Finally, and again like his father, Yongle would never forget the immense importance of reverence, an emotion he routinely evoked with publications and pronouncements, symbols and rituals, continuously dramatizing his intimate connection with the ancient powers and virtues of Buddhism, Taoism, and Confucianism.[9]

One can imagine that Zheng He, the admiral of the maritime fleet, was generally aware of his emperor's compelling needs. Moreover, as a leading member of the despised and distrusted eunuch service, Zheng He would have worried about the precariousness of his own status among the bureaucratic literati of the court. Castrated and enslaved as a boy after his Mongol father was taken prisoner while fighting the Chinese invaders of Yunnan, Zheng He had grown up demonstrating his exceptional talents as a leader in war and in peace, but he owed his political success and ultimately his life to the personal admiration and trust of Yongle.[10] The maritime expeditions were financially extravagant, however, and Zheng He may have astutely recognized that presenting the giraffes as ch'i-lin was one way to suggest the unique noneconomic value of the expeditions. Bringing ch'i-lin to China would glorify his own accomplishments, disarm his enemies at court, and flatter and support the emperor while spreading an anesthetizing fog of superstitious awe over the general public with a Ming version of "The Emperor's New Clothes."[11]

—

When the first giraffe arrived at the imperial court in 1414, the Board of Rites petitioned Emperor Yongle to accept a Memorial of Congratulation. Yongle modestly declined: "If the world is at peace, even without ch'i-lin there is nothing that hinders good government. Let congratulations be omitted."[12]

But when the second giraffe arrived the following year and was brought through the gates of the Imperial Zoological Gardens, the emperor himself attended the event, receiving obsequious prostrations from the dignitaries while accepting, along with a celestial horse (zebra) and a celestial stag (oryx), the blessed ch'i-lin. Yongle declared: "This event is due to the abundant virtue of the late Emperor, my father, and also to the assistance rendered me by my Ministers. That is why distant people arrive in uninterrupted succession. From now on it behooves Us even more than in the past to cling to virtue and it behooves you to remonstrate with Us about Our shortcomings."[13]

Zheng He's fifth expedition, which sailed two years later, reached the Arabian Peninsula, coming ashore at Aden, and then the eastern edge of Africa, stopping at Malindi, Mogadishu, and some other coastal trading settlements. Among the exotic treasures brought to the imperial court from that venture was an arkful of African animals, including antelopes, leopards, lions, oryxes, ostriches, rhinos, zebras—and more giraffes.

The final expedition ended in 1433, about a decade after Emperor Yongle's death. Yongle's successors were not so interested in the world outside, and so China's great Age of Exploration ended. By then the Chinese ruling class had become jaded about the exotic animals in the Imperial Gardens, so that only the first two giraffes brought from Malindi by way of Bengal during the fourth expedition were hailed as miraculous apparitions, the true embodiments of ch'i-lin, one of the four mythical beasts of Confucian tradition (along with the dragon, phoenix, and turtle), who had come to earth as evidence of a universal harmony induced by the unparalleled qualities of a great leader.

As Shen Tu, a poet and scholar of the Imperial Academy, wrote in his preface to a poem dedicated to the emperor:

> All the creatures that spell good fortune arrive. In the ninth month of the year *chia-wu* of the Yongle period, a ch'i-lin came from the country of Bengal and was formally presented as tribute to the Court. The ministers and the people all gathered to gaze at it and their joy knows no end. I, your servant, have heard that, when a Sage possesses the virtue of the utmost benevolence so that he illuminates the darkest places, then a ch'i-lin appears. This shows that Your Majesty's virtue equals that of Heaven; its merciful blessings have spread far and wide so that its harmonious vapours have emanated a ch'i-lin, as an endless bliss to the state for a myriad myriad years.[14]

Shen Tu declared himself a lowly servant of the emperor and, as such, wished to join the admiring throng. Beholding such an omen of good fortune as the ch'i-lin, he humbly lowered himself one hundred times while knocking his head to the ground in order to present an honorific hymn concerning the glories of the emperor and the unicorn, this "manifestation of the divine spirit," this very ch'i-lin.

This grand ch'i-lin who, combining a deer's body and the tail of an ox, stands fifteen feet tall, possesses a "fleshy boneless horn," and is colored "with luminous spots like a red cloud or a purple mist."

This gentle ch'i-lin who, anxiously examining the ground he walks on, is careful to avoid stepping on any living creature.

This harmonious ch'i-lin who, walking in such a stately manner, "observes a rhythm" for each movement he makes, while producing, with his "harmonious voice," the pleasing sounds of a bell or a "musical tube."

This glorious and benevolent ch'i-lin who, with his wondrous presence, magnifies the glories and benevolence of the Son of Heaven himself: "the Sacred Emperor who excels both in literary and military virtues," the one "who has succeeded to the Precious Throne and has accomplished Perfect Order and imitated the Ancients!"[15]

This chapter tells of the Chinese oceanic expeditions in the fifteenth century and the giraffes brought back and presented to the Ming emperor Yongle as "unicorns," based on an ancient Confucian tradition describing the unicorn as one of four magical animals. How could anyone mistake a giraffe for a unicorn?

The set below opens with photographs showing giraffe middle (median) horns: important if you're going to be mistaken for a magical unicorn. Also important was the fact that Confucian legend described the unicorn's horn as being covered with skin or hair, as are the horns of giraffes. Finally, Confucian legend stressed that the unicorn was a remarkably serene and gentle animal, blessed by nature or providence. These photographs suggest some of the steady grace and quiet gentleness of giraffes.

The Chinese seem to have brought home reticulated giraffes from East Africa. This set of photos likewise concentrates entirely on individuals of the reticulated group in the Samburu National Reserve, Kenya. Some of these giraffes have extra bony protuberances, or "horns," behind their ears and elsewhere.

ZARAFAS

FOLLOWING THE GREEK AND ROMAN HABIT, Arabic authors of the Middle Ages continued to identify giraffes as camelopards, but Arabic speakers began applying a word that sounded like *zurafa* or *zarafa*. The word, according to one early commentator, came from a linguistic root meaning "assembly," in reference to the idea that this animal was an assemblage of parts of different animals. Another Arabic scholar insisted that it derived from the Ethiopian *zarat,* meaning "thin" or "slender."[1]

Rome and the classical world had depended on Egypt as the sole gateway to deeper Africa and thus the sole source of giraffes, but after the disintegration of the Roman Empire in AD 476 and the subsequent expansion of Islamic civilizations that would finally control three-quarters of the Mediterranean, these animals were typically destined to become the private property of Islamic nabobs in the Middle East, who may have had occasion to speak among themselves about the rare, strangely formed, and remarkably gentle animals called zarafas.

The giraffes arrived, as before, from lands along the Nile to the south of Egypt—a part of Africa the Romans called Nubia, which included what the Ptolemies knew as the kingdom of Meroë. The Nubians, like the earlier Meroites, were occasionally acquiring live giraffes in trade with hunters and dealers from farther south. But the movement of giraffes into Egypt became more routine by 652, after Islamic forces took over Egypt and the emir Sa'd Ibn Abi Sarh proceeded to conquer Nubia. The emir settled on a treaty with Nubia that required the annual tribute of four hundred slaves, numerous camels, a pair of elephants, and a pair of giraffes. Later on the extortion was moderated so that the Nubians were sending three hundred slaves and a single giraffe annually. But the same steady traffic—in captive humans and animals—moving from sub-Saharan Africa through Nubia and into Egypt would continue for more than two centuries.[2]

Islamic cultures had by then created a powerful monopoly on most trade out of the northern end of Africa. Christian Europeans were excluded from participating in the profitable trans-Saharan traffic in gold and slaves, for instance; and, because the Nile River trade was likewise limited and blocked, Europeans were also kept from trading in—and ultimately kept ignorant about—giraffes.

By the time Saint Isidore, Bishop of Seville (560 to 636), was writing his *De Natura Rerum,* camelopards were obscure enough in the West that he accidentally combined them with chameleons, thereby producing a beastie he dubbed *chameleonpardus,* which resembled a cross between a camel and a lion—with a leopard's white spots, a horse's neck, the feet of an ox, and a coat that "changes to the colours which it sees by a very easy conversion."[3]

But perhaps the primary evidence of European ignorance about giraffes is found in the pages of a book known, in modern English, as *The Book of Beasts.* This strange, serious work was based on a core text originally written by someone who probably lived in Egypt between the second and fifth centuries AD. The anonymous author was influenced by the zoological writings of the ancients—Herodotus, Aristotle, Pliny, and the like—but the book became such an unbridled success that it spread, during the next few centuries, across the Middle East and Europe. Passed from monastery to monastery, it was copied entirely by hand and translated from the original Greek into Syrian, Armenian, Ethiopian, Latin, Arabic, Anglo-Saxon, Icelandic, Spanish, Italian, Provençal, and several dialects of German and the Romance languages. *The Book of Beasts* was probably second only to the Bible in being the most widely copied, distributed, and read book of pre-Gutenberg Europe.[4]

In fact, *The Book of Beasts* was the closest thing medieval readers had to a zoological encyclopedia. Or was it more a Wikipedia? During the process of making new copies and new translations, successive authorities with some knowledge, real or imagined, would simply add their own particular pieces of information, so that, for example, an early Greek version of the book references the natural histories of 49 animals, while the twelfth-century Latin version provides the natural history of some 110 to 140 animals. If we judge this material by contemporary standards, "natural history" is itself a generous expression. The book is profoundly flawed by the lack of any discipline for sorting truth from falsehood, while the background cultural presumption—that God imbued all of nonhuman life with a moral symbology for the edification of wayward humans—means that each animal is presented as a moral type or symbol.

The result is a strange mix of fragmentary knowledge, Christian moralizing, and audacious myth that strives to introduce a medieval reader to the possible lives and meaning of many beasts (bears, dogs, horses, lions, mice, and so on) and several of their lesser-known brethren, such as the unicorn, the griffin, and the sphinx. Then there is the monoceros: "a monster with a horrible howl, with a horse-like body, with feet like an elephant, with a tail like a stag's." Sirens, we learn, "are deadly creatures who are made like humans from the head to the navel, while their lower parts down to the feet are winged."[5]

The mysterious bonnacon has a bull's head and a body resembling a horse's mane, and its horns are so convoluted that they are incapable of inflicting damage, thereby forcing this creature to turn the other cheek: "For when he turns to run away he emits a fart with the contents of his large intestine which covers three acres. And any tree that it reaches catches fire. Thus he drives away his pursuers with noxious excrement."[6]

With such a splendid presentation of animals known and unknown, ordinary and fantastic, one has to wonder, where is the camelopard, that marvelous chimera the Greeks named and were once so familiar with? In fact, giraffes are simply absent from *The Book of Beasts,* as they must surely have been from Europe and from European zoological knowledge of the time.

Absent from Europe, that is, until the reign of Frederick II (1194–1259), who was king of Sicily from the age of three and, after 1220, the Holy Roman emperor.

Frederick was physically unprepossessing, at least according to one Islamic commentator, who described him as bald, nearsighted, and not worth 200 dirhams if sold in the slave market.[7] But he was learned, ambitious, and powerful. He spoke six languages. He patronized the arts and sciences. He promoted learning in an age of ignorance, skepticism in a time of dogmatism, and tolerance in an era of rabid intolerance. For such unusual and unpopular virtues, along with a reluctance to bring his army into the Fifth and Sixth Crusades and, later, his engagement in various maneuvers, plots, and battles involving Popes Honorius III, Gregory IX, and Innocent IV, Frederick was excommunicated four times. He was additionally identified as the Antichrist by Pope Gregory IX and posthumously placed by Dante in the sixth circle of hell: a bleak, dim, noxious section reserved for the tormented souls of notorious heretics.[8]

Frederick II, in short, was unfavorably reviewed by his peers, but he earned a special enmity during the Sixth Crusade by negotiating personally, as a splendidly treated guest, with the Egyptian sultan al-Kamil. After five months of positive diplomacy, the Holy Roman emperor and the Islamic sultan concluded a truce that handed over Jerusalem, Nazareth, and Bethlehem to the Crusaders while keeping the Dome of the Rock and the al-Aqsa mosque under Islamic control. Frederick then had himself crowned king of Jerusalem, thereby marking a peaceful conclusion to the Sixth Crusade. Pope Gregory IX expressed the concern that Frederick had already been excommunicated, which theologically invalidated his participation in the Sixth Crusade altogether. Other crusaders considered his bloodless treaty a self-serving betrayal of their sacred, sanguine cause. In any event, it was Frederick's relationship—likely enough, a genuine friendship—with the sultan al-Kamil that solidified his great notoriety while simultaneously prompting the wondrous gift of a single live giraffe to be shipped from Cairo, Egypt, to Palermo, Italy, supplementing Frederick's existing menagerie at the Palermo court. That giraffe was Europe's first since the days of the Roman Empire.

With the animal came the word. Starting in the thirteenth century—probably as a linguistic accompaniment to Frederick II's gift from the sultan—Europeans gradually became familiar with the Arabic sound *zarafa* and its echoing variants.[9] *Azorafa* appeared in old Spanish, eventually to become *girafa* in modern Spanish and Portuguese. Squeezed into old French, the Arabic word became *orafle* or *girafle* before, eventually, turning into *girafe* for modern French.[10]

—

Around two hundred fifty years later, a second zarafa sent across the Mediterranean as gift or exchange from an Islamic potentate to his European counterpart would also have a significant impact on European knowledge and perception of the species.

The Medici giraffe, as she is sometimes remembered, arrived in Florence on November 11, 1487, swept into town along with a lion and a horse, a scattering of fat-tailed sheep and long-eared goats, and the more exotic escort of several white-robed, blue-eyed Mamluks and their darker Moorish comrades who—bearing bags and trunks full of precious aromatics, cosmetics, ceramics, textiles, and the like—presented elaborate greetings from the Egyptian sultan Qa'itbay himself. As imagined by Marina Belozerskaya in her book *The Medici Giraffe,* the curious citizens gathered on the streets of Florence that day must have been left speechless by the sight of this rarest of animals: "at once bizarre and beautiful, muscular and dainty, moving both its slender right legs in one stride and its left ones in the next."[11] The arrival of this quadrupedal astonishment was a triumph for Lorenzo de Medici, and it happened not as the consequence of a love for animals or the collector's passion for novelty, but rather as the result of an ambitious and possibly desperate man's craving for power and status.

Lorenzo de Medici was born into the wealthiest family of Florence. He was physically strong, intellectually talented, and he also—as befitted a Medici—possessed a powerful sense of his own personal drama. The caved and crooked nose, the jutting jaw, and those "small hooded eyes" may have made him look, in Belozerskaya's assessment, "very much like a thug."[12] But for anyone as rich and prominent as Lorenzo de Medici in Renaissance Italy, a bit of thuggery would have been among the most important of personal qualities.

Less than a decade earlier, for example, and following a conflict over money with Pope Sixtus IV, Lorenzo and his brother Giuliano were set upon by a cabal of assassins in the great cathedral of Florence. The plot originated with Pope Sixtus and

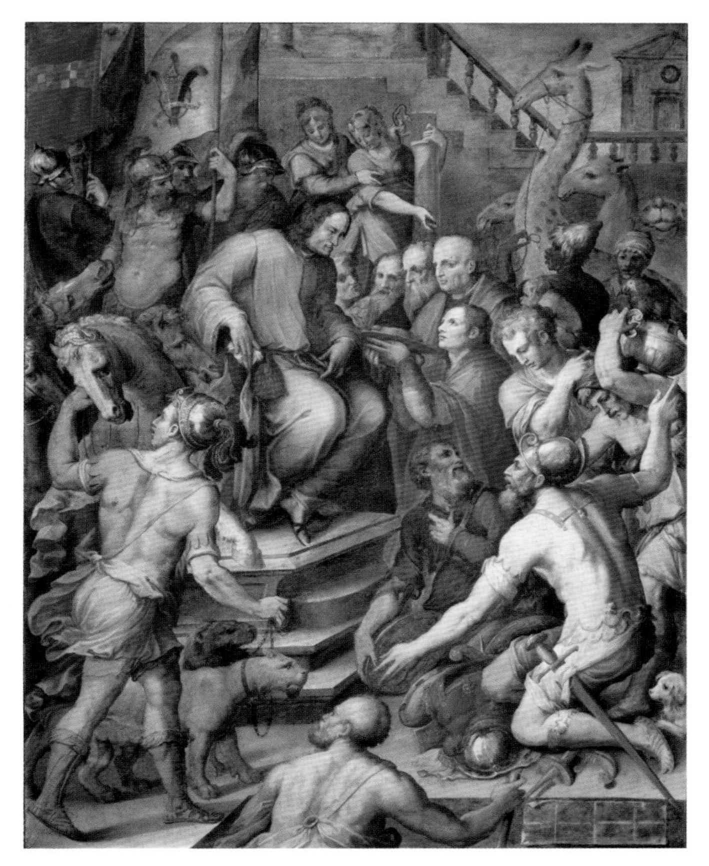

Giorgio Vasari (1511–1574). *Homage of All the Peoples to Lorenzo de Medici*. Sala di Lorenzo, Palazzo Vecchio, Florence. Photo: Scala / Art Resource, NY.

Archbishop Francesco Salviati of Pisa who, joining forces with two other rich and well-placed conspirators, engaged the services of a pair of loyal priests and a number of allies and mercenaries. After a ruse that brought both Medici brothers into the cathedral for high mass, the assassins surrounded the two men, drew their daggers, and went to work. Giuliano de Medici was killed outright, his blood flowing onto the cathedral floor from nineteen separate wounds. Lorenzo, after being stabbed in the neck by the two priests, defended himself with a hastily-drawn short sword well enough to escape into the north sacristy where, barricading himself behind a pair of massive bronze-clad doors, he waited for help to arrive.

Pope Sixtus, having failed with assassination, chose next to try excommunication, and that spiritual blow soon translated into a political one as other Italian potentates, sensing weakness, moved in to threaten Florence and extirpate the Medicis. Ultimately, Lorenzo handled the political crisis brilliantly, negotiating critical alliances while elevating the family's status and buttressing his own prestige and power in two additional ways.

First, he had his son Giovanni made a cardinal. This ascension became possible after Sixtus IV died in 1484, to be replaced by the more amenable Innocent VIII. The price for a cardinal's scarlet was nevertheless steep: not only an enormous loan to the pope and donation to the church, but also the marriage of Lorenzo's thirteen-year-old daughter to Innocent's thirty-seven-year-old illegitimate son, a repulsive drunkard and gambler. But for Lorenzo, at least, the exchange must have been worth it. In 1489, Giovanni was finally given the coveted position (although, since Giovanni was only thirteen years old at the time, he had to wait three more years before openly donning the cardinal's robe and miter). This, as well as the marriage of his young daughter to the pope's wayward son, eventually led to a far more direct connection between the Medicis and the Vatican: Giovanni became Pope Leo X in 1513, and a decade later the illegitimate son of Lorenzo's assassinated brother Giuliano was elected Pope Clement VII.

Lorenzo's second bold move was to acquire that giraffe from Sultan Qa'itbay of Egypt. Getting such an animal turned out to be approximately as expensive and difficult as having a son made cardinal, and its benefits were, likewise, potentially great yet impossible to measure and hard to predict. Certainly these creatures were, for Europeans of the late fifteenth century, very strange and rare indeed: a prize fit for a prince or a great dictator, which was how Lorenzo, born the son of a merchant, wished to present himself.

He already had a number of large beasts in his private menagerie, including leopards, lions, tigers, bears, and elephants.[13] A giraffe would have been a magnificent addition to an already splendid collection. But according to Belozerskaya, the ambitious Florentine had an even more dramatic sense of the value of that giraffe. Lorenzo, she argues, had begun to appreciate some interesting parallels between himself and Julius Caesar. More particularly, the former had been provoked by reading "with great interest" historical accounts of the lat-

I have seen [the giraffe] take sweetly from the hand of very young little girls bread, grass, fruits and onions which it eats with much voraciousness. I have also seen it raise its head up to those onlookers offering to it from their windows, because its head reaches as high as eleven feet, thus seeing it from afar the people think that they are looking at a tower rather than an animal. **—ANTONIO COSTANZO, 1486**

ter's 46 BC triumph in Rome, which featured, among so many other natural and unnatural wonders, the first giraffe ever to appear in Europe.[14]

Certainly Lorenzo's grandfather, Cosimo de Medici, understood that the difference between a rich merchant and a powerful prince was, as much as anything, a matter of style. He thus built and lavishly furnished the Medici Palace in Florence; and he also at one time planned an animal combat in the Piazza della Signoria that was meant to revive, albeit on a smaller scale, the notorious animal shows of ancient Rome.

Lorenzo was ten years old when the event was organized, and it was occasioned by the visit of Pope Pius II and Galeazzo Maria Sforza, son of the duke of Milan. What could a mere merchant, even a very wealthy one, possibly do to entertain and impress such powerful, princely men? Cosimo tried to answer that question by closing down the Piazza della Signoria and directing there the construction of an arena and grandstands.

The city of Florence took great pride in its pride of twenty-six lions, who were funneled into the arena once it was filled with enough potential victims for the gory drama to proceed: four bulls, a pair of horses, a couple of buffalos, some goats, a cow and calf, and a wild boar. The lions raced in, paced about, investigated. They roared. They sniffed. One of the big cats leaped onto a horse, and then, overcome perhaps by the puzzle of what to do next, she leaped off. The other twenty-five cats, having come and seen, chose not to conquer. They were already perfectly well fed and, in addition, had never been properly schooled in the origins of meat. Soon losing interest in their new surroundings, the bored creatures lay down, closed their eyes, and fell into a feline oblivion.

The pope was not amused. Ten-year-old Lorenzo, however, learned from the experience. Of course, it would not be possible for the grown-up Lorenzo to repeat Julius Caesar's great triumph in Rome, just as Cosimo had never thought to equal, at his makeshift arena in Florence, the enormous and bloody spectacles of ancient Rome. It might be enough merely to evoke the past in a certain subtle fashion. A passing reflection of the glory of ancient Rome should be enough. Hence the giraffe, who did indeed provoke a great sensation upon her arrival in Florence and was, over time, glamorously represented in word and image, with the latter including impressive paintings done by Botticini, Vasari, and Bacchiacca. The poet Antonio Costanzo was most impressed by the giraffe's astonishing gentleness. As he wrote in a letter to a friend in 1486, she had "neither ferocity nor cruelty" but rather was "so sweet," adding that he had seen her ambling freely through the streets.[15]

Exactly how the sultan of Egypt was persuaded to send Lorenzo that gentle animal in the first place is another story, the very short version of which is that Lorenzo used his influence with the royals of France to shift a prisoner—the brother of Qa'itbay's mortal enemy—from one European prison to another. More significantly, it appears that in arranging this peculiar move on the chessboard of real life, Lorenzo was forced to promise his anticipated giraffe to Anne de Beaujeu, the daughter of France's King Louis XI and, after 1483, the sister of King Charles VIII.

Referring directly to that "promise" in a poignant note written in the spring of 1489, Anne de Beaujeu begged Lorenzo to "deliver the animal to me and send it this way, so that you may understand the affection that I have for it, for this is the beast of the world that I have the greatest desire to see."[16] Lorenzo, unhappily, failed to satisfy that desire, apparently because the giraffe too soon died, having fatally wedged her neck in an overhead beam of her barn.[17]

—

There was a world of difference between the near-miraculous arrival of a giraffe on European soil in Frederick II's time and the magnificent appearance of one in Lorenzo de Medici's. Until Frederick's creature arrived in the mid-thirteenth century, Europeans were not fantasizing about camelopards, nor had they begun to hear of *zarafa* or speak of *giraffes*. In Lorenzo's day, near the end of the fifteenth century, giraffes were still remarkably rare and precious, but at least they were spoken of and known to exist. At least they had not stepped directly out of the realm of fantasy.

Following closely upon the arrival of Frederick's giraffe, in fact, a few brave travelers had begun reporting home about a larger world that included these strange animals. Marco Polo, the Venetian merchant who in the year 1271 began a twenty-four-year journey with his father and uncle into Central Asia and China, referred to giraffes three times in his *Travels*. Although he mistakenly placed them on the islands of Zanzibar and Madagascar, Polo may have seen one or a few in captivity during his return voyage through the Middle East.[18]

A few decades later, Wilhelm von Bodensele returned from a journey to the Middle East and produced in 1336 a brief account of seeing in Cairo a giraffe, which he called a *geraffan*. Von Bodensele's text became one source for the semifictional, highly popular literary compilation *Travels of Sir John Maundeville of St. Albans* (1356), which includes the news of "a kynde of beast" in Arabia "that is a fayre beast, and he is hyer than a great courser of a stead, but his neck is nere XX cubits long, and his crop and his tail lyke a hart and he make loke over a high house."[19]

And in 1384, still a full century before the Medici giraffe came to Florence, a baker's dozen of curious Florentines also visited Cairo and then returned home, where two of them wrote interesting accounts of the animals displayed in the sultan's menagerie. For Florentine Giorgio Gucci, the giraffes and an elephant were "two kinds of wonderful beasts, wonderful for those who are not used to seeing such beasts"; but the giraffe was "a much newer thing to see," while the name *giraffe* was, to him, a complete novelty as well. The second Florentine, Simone Sigoli, considered the sultan's giraffes similar to ostriches, except for the absence of feathers and their horse-like tails and heads. But their legs did resemble those of a bird, while the horns actually resembled those of a castrated ram. This strange conglomeration was "really," Sigoli insisted, "a very deformed thing to see."[20]

The occasional appearance of live animals and various travelers' reports began, in turn, to influence the European authors of zoological compendia. In the thirteenth century, both Vincent de Beauvais (in *Speculum Naturale*) and Albertus Magnus (in *De Quadrupedibus*) briefly mentioned giraffes, with both referencing the animal owned by Frederick II.

Some two hundred years later, a camelopard was mentioned in the Dutch bestiary *Dialogues of Creatures Moralized*, with a woodcut illustration that, alas, does not begin to suggest much more than a cloven-hoofed horse.[21] *Dialogues of Creatures Moralized*, incidentally, appeared in 1480, the decade of the Medici giraffe—as did Bernhard von Breydenbach's *Peregrinations in the Holy Land* (1486), an illustrated account of travels from Venice to Mount Sinai. This work distinguishes itself by including what could be the first somewhat accurate illustration of a giraffe to appear in printed form: a woodcut entitled *Seraffa* done by von Breydenbach's traveling companion, artist Erhard Reuwich.[22]

A small number of other travel and natural history books of this sort, some more and some less accurately featuring giraffes, would follow in that century and the next. But what I find remarkable is that real giraffes did not. It is true that around the same time Lorenzo de Medici acquired his giraffe from Sultan Qa'itbay, at least three other powerful Italians (Ferante, Duke of Naples; Alphonso II, Duke of Calabria; Hercules I, Duke of Ferrara) also seem to have added giraffes to their own already impressive menageries. Other zarafas from the Islamic East: Was this a fashion? A fad? Surely, in any case, it could be construed as evidence of an explosion in princely wealth, an expansion of Mediterranean commerce, and the blossoming of personal and cultural exuberance that marks the Italian Renaissance.

So it is an odd thing: After Lorenzo's time, no more zarafas. No more giraffes sent as serene and gentle exchanges across the great divide—no more, that is to say, for another three and a half centuries.

Formerly the grand lords, whatever barbarians they may have been, rejoiced in having beasts of foreign countries presented to them. In the castle of Cairo we saw several of those which had been brought there from all parts of the world, among these the animal commonly called Zurnapa, by the ancient Romans Camelopardalis. This is a very beautiful beast of the gentlest possible disposition, almost like a lamb, and more amiable or sociable than any other wild animal. Its head is almost similar to that of a stag, save that it is not so large, and bears small, obtuse horns six inches long and covered with hair. There is a distinction between the male and the female inasmuch as the horns of the males are longer; for the rest, both sexes have large ears like a cow, a tongue like an ox and black, and lack teeth in the upper mandible. They have long, straight, and graceful necks and fine, round manes. **—PIERRE BELON, 1553**

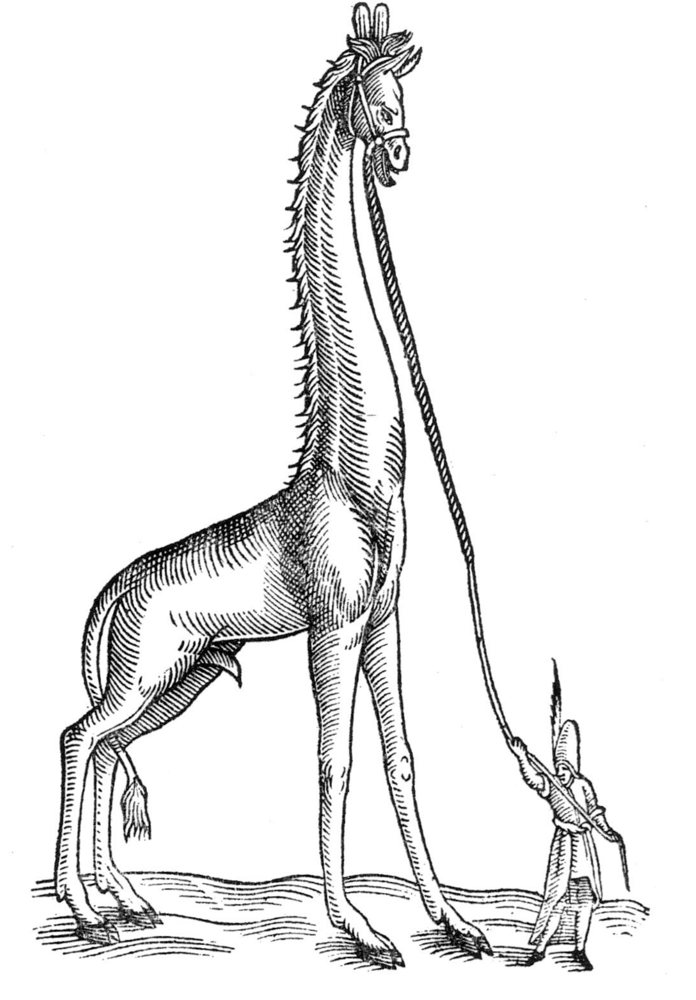

From Edward Topsell, *History of Four-footed Beasts*, artist Melchior Luorgius, 1607.

Following the Medici giraffe, then, European knowledge of giraffes depended entirely on those who traveled into Islamic territory. One such traveler, the French naturalist Pierre Belon, saw a giraffe in Cairo, whom he described accurately, in his *Observations de plusieurs singularitez et choses mémorables* (1553), as being "a very beautiful beast of the gentlest possible disposition," with "small, obtuse horns six inches long and covered with hair," as well as large ears, a black tongue, and no upper front teeth.[23]

But for those unable to manage such a difficult and expensive journey, a good imagination and a bit of bookish research would have to suffice. The English author Edward Topsell, for example, imaginatively portrayed in his *History of Four-Footed Beasts and Serpents* (1607) an animal with "two little horns growing on his head of the colour of iron, his eyes rowling and frowning, his mouth but small like a Hart's, his tongue is neer three feet long, and with that he will so speedily gather in his meat, that the eyes of a man will fail to behold his haste, and his neck diversely coloured is fifteen foot long, that he holdeth up above a camels and far above the proportion of his other parts."[24]

And where Belon's giraffe of 1553 was "a very beautiful beast," Topsell's 1607 creature was both beautiful and vain. When people come to look at them, the English author continued, these animals "willingly and of their own accord, turn themselves round as if it were of purpose to show their soft hairs, and beautiful colour, being as it were proud to ravish the eyes of the beholders."[25] We might think of Belon as attempt-

From Pierre Belon, *Les Observations de plusieurs singularitez*, 1553.

ing an objective, factual assessment. Topsell, by contrast, had no facts to be objective about—at least none based on his own observation.

Comparing the illustrations from these two books will clarify my point. Topsell illustrated his account with a woodcut by artist Melchior Luorgius, who (though it is said he once saw a giraffe at Constantinople) wildly misrepresented the creature's height. Standing next to a full-grown man in robe and turban, Luorgius's animal appears more than five times as tall as the man, which would make the animal over thirty feet tall. Belon's figure, though not accompanied by a human companion, looks to be of a realistic height for a giraffe. Belon had seen a living giraffe. Topsell never did.

—

From a scientific perspective, there is no substitute for direct and careful observation, yet Europeans searching for a true understanding of these animals would have no easy opportunities to look at them directly and carefully until they could break through the Islamic monopoly on trade with sub-Saharan Africa. Such a breakthrough actually began early in the fifteenth century, when Prince Henry, third child of the Portuguese King John I, sponsored a series of voyages that would open up the mysterious continent to European sailors and traders traveling south instead of east.

In 1415, at the age of 21, Prince Henry joined forces with his father and brothers to conquer Ceuta, an Islamic port on the northwestern tip of Africa just across from the Straits of Gibraltar. Ceuta had been a base for Barbary pirates who were raiding Portuguese coastal villages and selling the inhabitants into the African slave market. It was also an important terminus for the trans-Saharan trade in gold and slaves.

Henry, who was interested in the wealth of all that traffic coming in caravans out of the African desert, began to wonder how far south Islamic control extended. No one knew. Sailing ships on the Mediterranean were slow and heavy, but under Henry's guidance, Portuguese shipbuilders introduced a smaller, nimbler vessel, the caravel, that would effectively sail windward and, with the help of star charts, could be navigated on the open ocean without reference to shoreline markers. Sailing into the Atlantic with these improved ships and navigation aids, Portuguese sailors reached the island of Madeira in 1419 and by 1427 had passed the Azores. By 1488—a year after the Medici giraffe arrived in Florence—Bartolomeu Dias was rounding the southern tip of Africa and plowing a white wake into the Indian Ocean. These were primarily ventures of commerce and conquest, of course, concentrating on the riches to be gained from such precious commodities as gold and slaves.

This brave exploration and brutal exploitation also produced, during the next couple of centuries, the important side effect of expanding European knowledge about Africa: its people, geography, and animals, including giraffes. Beginning in 1624, a Portuguese missionary named Jeronimo Lobo entered the interior of East Africa from today's Mombasa, Kenya, and a year later explored the interior of Ethiopia by landing on the coast of the Red Sea.[26] His travel journals confirm that

Lobo may have been the first European to see a giraffe alive in sub-Saharan Africa. Finally published in 1728, the journals describe one as being among "the biggest of all the animals that we know, it is less bulky but more high than the elephant"—a quick sketch that was even more succinctly summarized in the entry in Samuel Johnson's English dictionary of 1755 as "an animal taller than the elephant, but not so thick."[27]

But giraffes would be found across Africa, more or less, and the early Dutch settlers in southern Africa, arriving by 1652, had within little more than a decade sighted them as far south as the Olifants River, according to a journal entry made by Pieter van Meerhoof and dated November 28, 1663: "Here we began to see the strange birds and flora . . . during our marching we saw two camels [giraffes], which animals have never before been seen or heard of."[28]

The discovery of giraffes so far south was not merely an interesting event. It was also a useful one, at least in terms of the developing scientific knowledge. European hunters in southern Africa would within the century begin sending detailed drawings, pelts, and even whole skeletons of these animals, thus making possible an increasingly accurate scientific understanding of what, or who, these animals are. The scientific study of giraffes, however, began far more simply, with the question, What kind of animal is this?

That was, of course, the very question the ancient Greeks had asked and then tried to answer with their theory of a chimerical hybrid. The first modern European to return to this original question was John Ray, a physician at London Medical College and the compiler of a systematic dictionary of animals published in 1668. Ray, having looked over the tantalizing if still insubstantial evidence, concluded that giraffes really did not belong in the same zoological group as camels. They were not a kind of camel. Rather, he thought, they belonged to the *Cervus* group, placing them into the fold of deer, sheep, and goats. Still, being in the *Cervus* group did not seem entirely satisfactory either, Ray concluded, since this "Camelopard seems to be a unique animal, called Giraffa by more recent writers."[29]

A generation after Ray, the Swedish botanist Karl Linnaeus promoted his own system whereby any plant or animal would be given a double name, in Latinized form, that would handily remind scientists of the interesting lines of anatomical similarity found in nearly all living things. In this double-name system, the first name recalled the larger group, the genus, to which the specimen belonged, while the second name recalled the unique species. Following John Ray's idea that giraffes, though odd enough, still belonged to the group including deer, sheep, and goats, Linnaeus settled on *Cervus* as the genus name. Then he added a species name that honored the old Greek term. Linnaeus therefore, in his *Systema Natura* of 1735, tagged giraffes as *Cervus camelopardalis*. Finally, though, the French zoologist Mathurin Jacques Brisson recognized—accurately, as it happens—that giraffes are anatomically distinct enough from all other living species to warrant their own genus, which he called *Giraffa*. By 1848, giraffes were officially recognized by scientists with the hybrid name *Giraffa camelopardalis*.[30]

—

Oh, yes, there were still zarafas. In the summer of 1827, three and a half centuries after the Medici giraffe came to Florence, an additional two living representatives of the species arrived in London and Paris. One was a gift to King George IV of England. The second was a gift to the recently restored monarch of France, Charles X. Both zarafas were sent by the Ottoman pasha of Egypt, Muhammad Ali, with the guidance and assistance of the French consul general in Egypt, Bernardino Drovetti.

Muhammad Ali, sometimes remembered as the father of modern Egypt, may also have been the biggest slave trader in history. Having conquered Nubia and Sudan in the early 1820s, the pasha was soon shipping down the Nile an estimated 50,000 black African slaves a year.[31] Drovetti, meanwhile, surely ranks among history's biggest grave robbers. He made a fortune dealing in antiquities: mummies and all the other dissolving remnants of Egypt's fascinating and glorious past. He also, on occasion, collected and sold live animals.

The specific occasion for this pair of ferocious men to send out those two gentle envoys was the Greek War of Independence. The Greek peninsula had been ruled by the Ottomans since 1453, but in 1821, Greek nationalists began an armed rebellion that would eventually lead to independence. The Ottoman sultan Mahmud II, based in Constantinople, discovered his own army to be incapable of controlling the rebellion, so he instructed his Egyptian viceroy, Muhammad Ali, to assist. Ali sent an army under the command of his son, Ibrahim, across the Mediterranean to Greece. They landed in February 1825 and did, soon enough, retake the peninsula and the city of Athens for the Ottomans. Of course, Ali smartly anticipated that this invasion would not be popular among many Europeans, and as a possibly subtle way of softening the blow, he instructed that the two giraffes be captured and sent to the rulers of England and France.

One might imagine that the gift of a mere animal would have little effect on the life-and-death decisions of great kings and their advisors, but these were not mere animals. Not only were they the first giraffes to reach Europe since the Italian Renaissance, they were also the first of their kind ever to appear in France and England. Moreover, Charles X had recently been requesting exotic animals to expand the collection of the royal menagerie in Paris. (And since he had lost his oldest brother, Louis XVI, to the guillotine in 1793, followed by a sister in 1794, Charles might have been constitutionally unsympathetic to messy rebellions and unruly revolutionaries.)

As detailed by author Michael Allin in his book *Zarafa* (1998), the pasha placed an order for two young giraffes in the summer of 1824, and in the fall of that year Arab hunters assigned to the task set out from the outpost of Sennar, in southern Sudan. A ten-day trek with horses and camels took them farther south and east into the savanna highlands, where they found giraffes. Giraffes can ordinarily run faster than horses but only in short bursts, so the hunters, working in relays, would chase down a mother until she was exhausted, then slash at her hamstrings with a sword. Crippled, she would collapse, whereupon the hunters finished her off, cut her up, and rounded up her baby.

In this way, the pasha's hunters acquired a good deal of meat and leather, along with two lost and frightened six-foot-tall infants, who were lashed onto camels and fed camels' milk for the next several days as the group returned to Sennar. At Sennar the animals were hoisted onto a felucca and floated for two hundred miles (322 kilometers) down the Blue Nile to the newly built slave garrison at Khartoum, where the Blue and White Niles join. They rested and grew during a sixteen months' stay in Khartoum before being loaded onto another felucca. This vessel had two lateen-rigged, sharply triangular sails and was up to forty feet long but light enough that, with the human and animal cargo offloaded, it could be pushed, pulled, and poled across the river's six cataracts. The human cargo would have been slaves. The animal cargo may have included monkeys along with one or two domestic cows assigned to produce milk for the giraffes.

It was a two-thousand-mile (3,219-kilometer) float from Khartoum to Alexandria, and during an extended period of recuperation in Alexandria, the two young animals were prepared for the next two-thousand-mile leg of their journey. It could have been at Alexandria, then, that someone provided them with protective amulets, each consisting of a few verses from the Koran tucked inside a carved wooden box and hung around their necks. The amulet must have been less effective for the England-bound giraffe, however, since she never really thrived. She remained in Alexandria until January and then, because of an extended delay in Malta, did not reach King George IV in London until the summer of 1827. Soon after, the unfortunate creature had to be rigged into a prosthetic harness that kept her upright. She died within the year.

The giraffe bound for France, however, adapted well to her new circumstances. She spent three weeks sailing west on the Mediterranean—standing in the hold of a two-masted brigantine alongside three milk cows, two cavalry horses, and a pair of African antelopes, while raising her neck and head through a hole cut into the deck so that she could appreciate the fresh air and bright light outside.

The boat anchored in the French port of Marseille, and after a week's quarantine, on the last day of October 1826, this gift for King Charles X walked down the gangplank and was taken to her winter quarters in Marseille. The French consul general in Egypt, Bernardino Drovetti, had carefully arranged most of the details for this challenging trip, and indeed Drovetti's Arabic syce and his Sudanese servant, two men remembered now only by the first names Hassan and Atir, accompanied the precious animal. But possibly her closest companions were the three Egyptian milk cows who had shared the ship's hold with her and who continued, since she was never weaned, to add substantial milk to her diet.

The giraffe had learned to trust and therefore follow the

Nicolas Hüet the Younger (c. 1770–1827). *Study of the Giraffe Given to Charles X by the Viceroy of Egypt*. 1827. Watercolor. Purchased on the Sunny Crawford von Bülow Fund, 1978. Photo: Pierpont Morgan Library / Art Resource, NY.

cows, so when it was time to begin the 550-mile (885-kilometer) trip from Marseille to Paris in the spring of 1827, the cows preceded the giraffe, while the giraffe, wearing a custom-cut raincoat, dragged three long ropes attached to three earnest men. They walked all the way, joined by the distinguished zoologist Étienne Geoffroy Saint-Hilaire and protected fore and aft from curious pedestrians and distracting traffic by mounted gendarmes. It was a lonely procession at first, but it grew bigger. By the time it reached the town of Lyons, some thirty thousand people turned out to see the animal. The giraffe and her attendants passed through Paris on July 9 in a great parade, escorted by the royal cavalry, which took them to the royal palace at Saint-Cloud for an official rendezvous with Charles X. Then it was back into Paris, where the animal was quartered in a greenhouse at the Jardin du Roi and allowed visitors: some one hundred thousand over two months, which was around an eighth of the city's population.

The pasha's gift had the kind of electrifying effect an alien dropped in from outer space might produce, except that she was simultaneously alien and, somehow, splendidly familiar. Her sinuous beauty, her aloof serenity, her tentative gentleness seemed to inspire—for a season, anyway—all of France. Women piled their hair into towers *à la girafe*. Men donned *girafique* hats and ties. "Zarafamania," writes author Allin, "was everywhere—in textiles and wallpaper, crockery and knickknacks, soap, furniture, topiary—anywhere her distinctive spots or long-necked shape could be employed."[32]

It might seem that this zarafa had provided just the public relations effect originally intended by Muhammad Ali. In fact, on the very day the animal and her entourage entered Lyons, June 5, the Egyptian army commanded by Ali's son had finally taken Athens on behalf of the Ottoman sultan in Constantinople.

Within a month, however, the British, French, and Russians had combined to demand that the sultan withdraw those forces. By October 20, the European allied fleet had sailed into the Greek Bay of Navarino and annihilated the entire Egyptian and Ottoman fleet. It was the end of Ottoman control in the Peloponnese and the beginning of Greek independence, which was finally achieved in May 1832. In France, around the same time, royalty had once again started to drift out of fashion, so that Charles X was forced to abdicate in 1830. Meanwhile, inside the Jardin du Roi and serenely ignorant of princes and kings, sultans and pashas—and of the madness, greed, cruelty, and violence infecting the fragile lives of those strange and short and ever-chattering two-legged creatures who passed into and out of her vision—the pasha's zarafa gazed and ate, gazed and ate. She was joined by a companion zarafa in 1839 and died on January 12, 1845.

The giraffes imported by the Ptolemaic Egyptians, then again imported by the later Islamic rulers of Egypt and given as tribute gifts, were taken from southern Sudan and Ethiopia. They would have been members of the Kordofan and Nubian subspecies (or, as some geneticists would have it, species). Kordofan giraffes are today very rare and could be extinct, while the Nubian giraffes have dwindled to an estimated 160 individuals. The zarafas, in any case, may have come from arid plains that included brush and woodland but also areas of desert or near desert. Certainly during their passage down the Nile and into Egypt they would have crossed through that great northern barrier of mountains and desert we call the Sahara. These photographs of the desert-adapted Angolan giraffes living in northwestern Namibia evoke the vanished zarafas and their world.

In the opening photograph for this chapter, some Angolan giraffes, seen from afar, are illuminated by a sudden burst of sunlight. The mountain would ordinarily be rust red, covered as it is by rocks with a high ferrous content; the green is from sparse and spotty vegetation that appears transiently after the wet season. The images below show the illuminated giraffes as they gather, then turn and run. The sequence then shifts to a different group of giraffes in an even more severe environment (in reality the shift covered several hours and perhaps one to two hundred miles off road and into the trackless desert).

Karl and I went to Namibia at the end of the rains and the beginning of the long dry season. It was hot. Some of the images in this sequence show the effects of photographing unusually shy animals from a very long distance via telescopic lens while a rising heat turns the air into a distorting lens. Others are marked by the harsh desert light. The giraffes in this barren world can survive without drinking water, gathering sufficient moisture from the bushes and trees that grow along seasonally dry riverbeds and elsewhere.

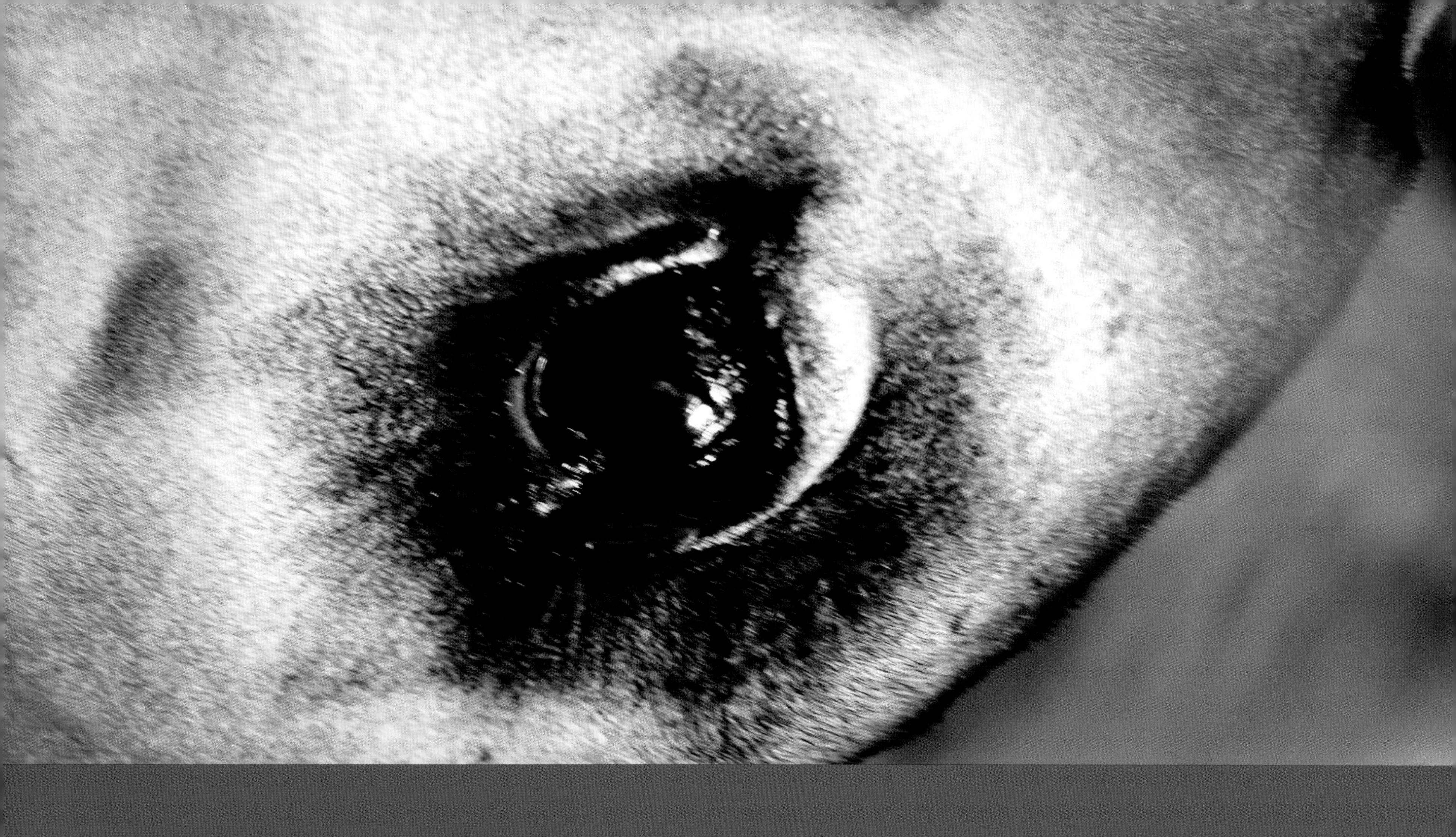

GIRAFFIDS

IN THE ITURI FOREST of the northeastern Democratic Republic of the Congo lives an extraordinary animal called the okapi. I sometimes think of okapis as forest giraffes. Although giraffes stand alone and lonely—the sole members of their own genus *Giraffa*—okapis linger not far outside that genus. Giraffes and okapis happen to be each other's closest living relatives, even though at first glance they may seem like very, very different creatures indeed—okapis being short-necked, shadow-colored, and lurking in a swirl of shadowed rainforest; giraffes being long-necked, sun-colored, and living in sun-blasted savanna and woodland. Okapis are also much rarer than giraffes, and generally asocial. They are elusive enough that one can only guess how many remain. Perhaps several thousand. A more informed guess, for those okapis still living within the peripheries of the Ituri Forest's 5,300-square-mile (13,726-square-kilometer) Okapi Faunal Reserve, would be around 2,500 individuals.[1]

Rare. Asocial. Elusive. They are so elusive, in fact, that Karl and I probably would never have seen one had the Okapi Faunal Reserve not included right next to its official headquarters a conservation and breeding center that, when we were there, kept about a dozen of these animals in large fenced enclosures. Having dropped our bags in a guest house at the center, we quickly walked along a trail to the well-maintained enclosures for our first look.

To me, they looked like horses. They were about the size of horses. They seemed to have the general shape and structure of horses, with gleaming, dark chestnut coats and pale grayish faces. They had large, glistening eyes and abnormally long and dark tongues. A black-and-white zebra-style striping was wrapped around the high parts of their front and hind legs, while a similar zebra pattern fanned horizontally out from the hindquarters onto the haunches. From the rear, then, I saw a zebra, though one with mostly horizontal rather than vertical stripes: like highlights of water reflecting off dark leaves. From the front, I saw a shy dark horse, but one with very large ears and a long, bony face masked in a ghostly gray. I also saw cloven hooves.

—

The pygmies of Central Africa, who survive by hunting and gathering, by trading with cultural outsiders and occasionally working for them, are the world's great forest experts. In a small roadside village not far from the okapi center, Karl and I visited some Mbuti pygmies, a couple dozen of whom were gracious enough to take us on an afternoon's hunting expedition, done with long nets and iron-tipped spears and small brown dogs. After the hunt, some of the pygmies talked to us briefly about life and okapis.

Later on, in a smaller Mbuti village, we met an old man named Mayanimingi, who as a young man had worked in the original okapi capture program. Mayanimingi had a few thin spirals of hair making up a beard and a wide, bulbous nose. He was barefoot. He wore shorts and an oversized red sweatshirt and had an old red cloth wrapped around his head. He sat on a chair made of two stick triangles braced together with vines. He was short, as pygmies are, and very lean, and he seemed tired from age and the trials of his life. "La vie durée," as he memorably put it. Mostly, he gestured emphatically at us and conversed with our translator in a combination of Lingala and Swahili and fragments of French, which was then pressed into longer fragments of French for our benefit.

He did not know how old he was, Mayanimingi said, but he began working for a Portuguese man named Jean de Medina a long time ago, around 1952. (The original capture operation was begun near the start of the century by a European missionary and later taken over by an American anthropologist.) They trapped okapis by digging many pits in an area where they knew the animals were feeding. Obscured by leaves and branches, the pits were about two meters deep and long but comparatively narrow. They padded the bottoms of the pits with leaves and branches to keep any trapped okapis from injuring themselves. They would build a temporary enclosure right around the trapped animal and keep him there for three days before the move to a more permanent pen. Once inside that pen, the okapis were fed with leaves taken daily from several species of trees and bushes and vines, the fresh leaves being brought immediately to the pens and strung up, like clothes on a line, so that the captured okapis would be eating their usual foods at the usual level in their usual upward-reaching posture.

—

Possibly the first Westerner to hear about okapis, the English-American journalist and explorer Henry Morton Stanley, did so during conversations in 1888 with some Mbuti men from the Ituri Forest. The pygmies spoke of a strange and unusually large forest animal. Stanley was under the impression that the animal might be a kind of donkey, but a leaf-eating one living in a leafy kingdom. The explorer published his account of the conversation, and thus probably the first written reference to okapis, in his memoir, *In Darkest Africa* (1890). "The Wambutti," he wrote, "knew a donkey and called it 'Atti.' They say that they sometimes catch them in pits. What they can find to eat is a wonder. They eat leaves."[2]

Around the same time, an Englishman named Harry Johnston became interested in the cryptozoology of Central Africa's then unexplored forests. Johnston read Stanley's published account. More significantly, he had some extended conversations with a group of Mbuti pygmies who in the spring of 1900 were staying at his house in Entebbe, capital of the Ugandan Protectorate in East Africa.

Johnston was there as governor of the protectorate. The pygmies were there because a German showman had kidnapped them from their village in the Ituri Forest, intending to display them as live exhibits in the Paris Exhibition. Since the Ituri was by then part of the newly created Congo Free State, and since King Leopold II of Belgium claimed the Congo Free State as his own personal piece of Africa, Belgian soldiers chased the

German showman and his victims out of the Congo. They fled east into British-held Uganda, where Johnston was able to rescue the pygmies. Then, while he was preparing to escort them back to the Ituri Forest, the grateful Mbuti told the curious governor about an animal known as *o'api*. Johnston showed them a zebra skin and held up a picture of a donkey, and the pygmies agreed: The combination approximated an o'api.

Johnston organized an expedition to the Ituri Forest that July, returning the Mbuti to their homes while hoping to search for the *atti* or *o'api*—or, when more fully Anglicized, the *okapi*. Lieutenant Meura, the Belgian commander of Fort Mbeni, an outpost on the edge of the Ituri, confirmed that he had seen okapi carcasses and skins. Meura believed the animal was a horse of some kind, and he arranged for trackers to guide Johnston into the forest for several days. Johnston, unfortunately, was soon overwhelmed by the intense heat and humidity and dark oppressiveness of the place. As he later described the experience, "The atmosphere of the forest was almost unbreathable with its Turkish-bath heat, its reeking moisture, and its powerful decay of rotting vegetation. We seemed, in fact, to be transported back to Miocene times, to an age and a climate scarcely suitable for the modern type of real humanity."[3] At one point, his pygmy guides excitedly pointed out okapi tracks in the wet earth—but they were the tracks of a cloven-hoofed animal. Donkeys and horses are not cloven-hoofed, Johnston knew, and he suspiciously—or stupidly—refused to follow the tracks. Soon he and the rest of the group had succumbed to the high fevers and intense lassitude of malaria. They managed to return to Fort Mbeni where, aided by a cadre of troops and assistants and wrapped in a bitter disappointment, Johnston prepared for the difficult journey back to Entebbe.

Lieutenant Meura, meanwhile, had located some bits of okapi skin that were cut up and sewn into a pair of bandoliers, and he pressed those items into Johnston's hands. Apparently out of sympathy for Johnston's condition and disappointment, Meura also promised to find a complete okapi skin and forward it to Entebbe as soon as he could. Meura then died of blackwater fever, so he failed to fulfill the promise. His second-in-command, however, a Swede named Karl Eriksson, did; and by March of 1901, Johnston had received the full skin plus a pair of specimen skulls with a jawbone. In a note accompanying the package, Eriksson wrote that he had also sent hooves—cloven hooves—which did not reach Johnston and must have been lost in transit. Johnston recognized from the description of the hooves and a quick examination of the skin and skulls that this new species was not a close relative of horses or donkeys. No, this was something else entirely. What? The skin suggested not a long neck, like a giraffe's, but rather a back rising smoothly into the line of the neck—more like one of the extinct giraffe relatives already known from the fossil record. A member of the Giraffidae family: a giraffid.[4]

	GIRAFFES	OKAPIS
Kingdom:	Animalia	Animalia
Phylum:	Chordata	Chordata
Class:	Mammalia	Mammalia
Order:	Artiodactyla	Artiodactyla
Suborder:	Ruminantia	Ruminantia
Family:	Giraffidae	Giraffidae
Genus:	*Giraffa*	*Okapia*
Species:	*Giraffa camelopardalis*	*Okapia johnstoni*

But most compelling were the skulls, with their remarkable horns and teeth. Actually, only one of the okapi skulls had horns. The other did not. Nevertheless, Johnston recognized that the okapi's horns strongly resembled giraffe horns, while both were different from the keratin-sheathed horns of some other animals—cattle, for example. Indeed, some experts prefer to use the more precise term *ossicones* when referring to the horns of giraffes, okapis, and the ancestral line of extinct giraffids.

Both male and female giraffes are born with ossicones, although for females they remain less developed: they are generally flattened, tapered, and comparatively light. For males, they turn into a pair of weapons that are large, round, hard, and knobbed at the ends. At birth, the ossicones are about an inch (2.5 millimeters) long and consist of compressed cartilage, rather than the bone we associate with horns. The cartilage is not attached to the skull and is flattened at birth, yet it soon becomes erect, like a horn, while remaining covered with skin and wavy black hair. Within a week, the cartilage begins to ossify, or change into bone, and within two years it will be entirely bone, though still covered by a layer of skin. Eventually, the ossification spreads into the base, finally fusing the bony ossicone to the bone of the skull, which happens around four years for males and seven for females.

Ossicones grow and fuse to the skull on the parietal bones, located above or behind the eyes, and for that reason they are sometimes called "parietals" or "parietal horns," which is one way of distinguishing them from the other horns that may appear on a giraffe's skull. Most common is the median horn, mentioned earlier, which can appear as a pronounced bump or rounded projection in the middle of a giraffe's nose. Since this median horn is a direct outgrowth of bone, it superficially resembles a more ordinary horn, even though, like the ossicones, it remains covered by skin.

Just as the ossicones develop more fully among giraffe males, becoming weapons for male-against-male battles over mating access to females, so this median horn will also be more pronounced among males. Indeed, the fighting style of males (who hammer each other with their heads, using those long, powerful necks as immense levers) probably accounts for the pronounced sexual dimorphism in skull size and heft among giraffes. During a male's lifetime, the entire top of his skull continues to thicken and gain mass with expanding bone deposits. This thickening begins in the regions supporting the parietal and median horns; it expands to include much of the outer striking surface of the skull, to cover the surface blood vessels, and finally to produce a massive, armored skull that may have additional bony accretions behind the ears and elsewhere. An old male's skull, larger and thicker and irregularly accreted, may weigh as much as 13 kilograms (28–29 pounds), which is around three times the mass of a typical female skull.[5]

Among okapis, only the males have ossicones in the first place. They remain a comparatively small pair of unbranched projections looking rather like a young giraffe's, except that they taper to a point at the ends. Like giraffe ossicones, they are located on the parietal bone and covered with skin, although the skin does eventually slough away to expose the sharp bone at the tip. If we were to think of a male okapi's ossicones as weapons, then, we might imagine them to be the kind that stab rather than, as with the male giraffe, hammer.[6]

So while there are significant differences between the weaponry of okapis and giraffes, these two species are the only living mammals with ossicones rather than horns or antlers: a unique giraffid feature that was quickly recognized by Harry Johnston at his home in Entebbe, Uganda. At least as compelling as the ossicones, however, were the teeth on Johnston's newly received specimen skulls and jawbone. Taxonomists have always liked teeth, in part because teeth fossilize well. They are extremely hard, unlikely to be chewed or eaten by predators, and very resistant to erosion. Teeth also evolve in response to changes in food source and eating style, so they will generally tell you a good deal about the kind of animal possessing them: vegetarian, carnivore, omnivore, grazer, browser, and so on. And just as food sources and eating styles are predictably conservative or slow to change for any species, so tooth structure is likewise evolutionarily conservative. Teeth, therefore, remain a good index of relatedness across time.

Both giraffes and okapis share with other giraffids a tooth pattern that includes, in the lower front jaw, a curved prow of six incisors flanked by a pair of canines. The incisors and their flanking canines are projected outwards rather than more directly upwards. A large gap separates these lower front teeth from the back teeth, which are crescent-ridged grinding premolars and molars, tight rows of six on either side. Opposing this lower set of grinders is an equivalent upper set lodged in the skull, but at the front of the skull, the lower incisors and canines bite up against nothing more than an upper pad of thick, fibrous material. No upper incisors or canines at all.

This pattern, so far, repeats that of several other species. The defining feature of giraffid teeth appears once we consider more fully those two canines in the front of the lower jaw. We ordinarily think of canine teeth as the sharp rippers and tearers of the dental community, evolved in the service of carnivores or omnivores. The canines of giraffids, however, are round rather than sharply pointed. They look almost like incisors. They are shaped less like daggers and more like spoons, albeit double-lobed spoons. The double lobe is especially distinctive, and it provides this spoon with a notch in the middle. That remarkable notch assists—for both giraffes and okapis, and probably for the extinct ancestral giraffids as well (who also had lobed canines)—in a characteristic feeding style.

Johnston did not have an okapi tongue in his collection. If he had, he may have recognized yet another obvious similarity between giraffes and okapis: an exceptionally long and darkly pigmented tongue. Okapis can lick their eyelids or inside their ears with that tongue, and they use it in a prehensile fashion, as giraffes do. They wrap it around a leafy branch and draw it into the mouth, where, with a swiping motion of the head, the branch is passed through a comb of incisors and double-lobed canines, effectively stripping away any leaves, twigs, fruit, thorns, or other edible items.

Using the diplomatic pouch, Johnston sent his package of okapi specimen pieces, accompanied by an explanatory letter, to E. R. Lankester, director of the British Museum of Natural History in London. Lankester agreed that the okapis were giraffe-like. Indeed, he understood them to be members of the family of giraffids, though not immediately related to giraffes—cousins rather than siblings. Okapis were thus in 1901 given their own genus, *Okapia,* while the full species name would honor the man most fully associated with its discovery for science, making these remarkable animals officially *Okapia johnstoni.*

—

A few bones and some teeth provided the compelling evidence that enabled Harry Johnston and E. R. Lankester to complete their puzzle: demonstrating an anatomical similarity between okapis and giraffes significant enough to suggest an important family connection. When bones and teeth become mineralized and acquire the permanence of stone, as fossils, and thereby enter the groaning vaults of deep time, they can enable us to work this same puzzle—of anatomical similarity and family connection—far more expansively. Fossils, organized and compared well enough, provide that magical entry into the fourth dimension, where we can look across time as if flying across a rough and endless sea of it, contemplating, from that high and moving vantage, the shape-shifting history of things.

Based on the fossil evidence, we must now conclude that the giraffids—members of the Giraffidae family—emerged from a much larger group of mammals known as the artiodactyls (members of the order Artiodactyla) beginning around 55 to 60 million years ago.

"Artiodactyl" is a Latinized coinage that means, simply, "even-toed." We sometimes recognize artiodactyls as cloven-hoofed animals, but even-toed or (for giraffes) two-toed is a more accurate way to think about it. Evolution works not through instant invention but rather through gradual modification, and in the case of giraffids and other two-toed species, the standard mammalian frame—four projective limbs, each one terminating with five extending digits or dactyls—has been gradually modified in giraffids. Three of the digits (first, second, and fifth) are lengthened, drawn up and back, and then partially or fully fused together to become functionally part of the leg, while the third and fourth digits remain extended outward and have become dense and strong enough to support the weight of the body. What's the point of all that evolutionary manipulation? The changes produce a longer stride and thus a faster run. Meanwhile, protecting the ends of those two weight-bearing toes is a pair of dense and thickened toenails. What we think of as hooves, then, are modified toenails; and what we think of as the split hoof is actually a pair of side-by-side toenails that, though fused at the back, remain unfused in the front.

All giraffids are artiodactyls, or even-toed mammals. They are also ruminants, being members of the suborder Ruminantia, which, beginning around 40 million years ago, emerged as one distinctive line from the order Artiodactyla.[7]

The ruminants show signs of a specialized adaptation to eating high-cellulose plants. Their teeth have become higher-crowned, their molars broadened and also roughened with crescent-shaped cusps. Along with these distinctive teeth comes an even more distinctive digestive system that enables ruminants to draw nutrition from the cells of cellulose-dense plants more fully and efficiently. This new and improved kind of digestion, however, requires a ruminant's four-chambered stomach. An initial chewing of plant matter in the mouth mixes it with saliva and passes it down to the first chamber, the rumen, where the roughly chewed vegetation is separated into solid and liquid. The liquid is passed to the second chamber, while discrete chunks of the solid material (each chunk a bolus) are returned to the mouth for another, more thorough chewing (ruminating) that further processes the material before it is passed down to the third and, from there, to the fourth chamber and finally to around 280 feet (85 meters) of intestine.[8] That four-part system, that gastric roller-coaster of downs and ups, ins and arounds, may seem to us single-stomached humans like an illogically complex way of doing things. But then how many humans can survive on raw leaves and twigs and thorns?

Just as the suborder Ruminantia represents one branch that emerged from the many-branched order Artiodactyla, so the Ruminantia would divide, at around 18 to 21 million years ago, into five distinct families that include the Giraffidae family—the giraffids. There were, at various times, a lot of giraffid types, and they spread broadly into Europe, Asia, and Africa. All told, the family branched into one to two dozen genera, including the pair of genera I have already referred to: *Okapia* and *Giraffa,* each with some extinct species and one still-surviving species (*Okapia johnstoni* and *Giraffa camelopardalis*). All the other giraffid genera and, of course, all their species are now extinct.

ORDER: Artiodactyla
SUBORDER: Ruminantia
FAMILY: Giraffidae
INCLUDES THE GENERA:
Birgerbohlinia
Bohlinia
Bramatherium
Decennatherium
Giraffa
  Includes the surviving species *Giraffa camelopardalis*
Giraffokeryx
Helladotherium
Honanotherium
Macedonitherium
Mitilanotherium
Okapia
  Includes the surviving species *Okapia johnstoni*
Palaeogiraffa
Palaeotragus
Propalaeomeryx
Samotherium
Shanistherium
Sivatherium

The giraffids are most obviously marked from the rest of the artiodactyl ruminants by the features noticed by Harry Johnston and E. R. Lankester in 1901: skin-covered horns, or what Lankester called ossicones, and the double-lobed canine teeth. Many members of this large and complex family, however, looked very, very different from the two species remaining today.

Members of the *Sivatherium* genus, for example, were about the size of elephants. They had heavy legs, short necks, and broad skulls, and the males sprouted massive and complex ossicones looking rather like a trophy rack of elk's horns. They originated in Asia and eventually evolved into two species, Asian and African, with the Asian one (*Sivatherium giganteum*) becoming extinct in the early Pleistocene and the African one (*Sivatherium maurusium*) apparently surviving in North Africa until about 3,500 years ago.[9] The impressive elk-like ossicones of *Sivatherium* were anchored in the skull well behind the eyes, while a second pair of much smaller and simpler ossicones was located more directly above the eyes. Most of the extinct giraffids had larger and more numerous ossicones than the two surviving species; *Bramatherium* and *Giraffokeryx* species shared with *Sivatherium* the general pattern of four ossicones.[10] Members of the *Giraffokeryx* genus were medium sized, and recent artistic reconstructions show an animal looking rather like an okapi, though with a somewhat longer neck.[11]

Fossils of the long-legged and long-necked *Bohlinia* genus resemble modern giraffes enough that some experts consider that genus directly ancestral to the *Giraffa* genus. *Bohlinia,* then, may have begat *Giraffa;* and the earliest members of *Giraffa* migrated east, eventually splitting and begetting around four or five Asian species. Others of the genus moved to the south and west, crossing the Afro-Arabian landmass and passing into Ethiopia around 7 million years ago. These animals continued moving south and west and over time split into about five African species.

The approximately ten species of *Giraffa* varied in average size, general shape, and location and orientation of their ossicones, but they were all still close enough in appearance to the surviving contemporary species, *Giraffa camelopardalis,* to be recognized as members of the same genus. The first fossils that clearly mark the one survivor began to appear in formations that originated around 1 million years ago, but the single most obvious characteristic of that animal—the long neck—goes back at least 7 million years. In fact, an elongated neck within the larger giraffid family may have appeared even earlier, considering that remains of the long-necked *Bohlinia* have been recovered from geological formations dated at 7 to 9 million years.[12]

These photographs of okapis, the only living relatives of giraffes, were taken at the Okapi Conservation and Breeding Center in the Democratic Republic of the Congo. The image that heads the chapter, a semi-abstract close-up profile of an okapi, is followed (below) by a portrait in which a piece of green vegetation, close to the camera, produces a verdant overtone that may suggest some of the elusiveness of okapis within their forest refuge. Note that the photographer seems to be reflected in the subject's eye.

The second image in the sequence below provides an excellent demonstration of the lovely chestnut overtones of this animal's pelage when seen in a certain light, while the third, of a female chewing at a minor irritation on her haunch, shows the horizontal zebra-like striping at the rear, which may provide useful camouflage when the okapi is standing still within a dark and wet forest. The fourth image gives us a sense of the unusual coloration and length of an okapi's tongue, interestingly reminiscent of a giraffe's tongue, while the fifth and final photograph—a full head-and-body portrait of a young adult male—provides some sense of an individual with personality who may not appreciate having his picture taken. Even a fully mature male okapi's horns (ossicones) will look small and sharp when compared to a giraffe's horns.

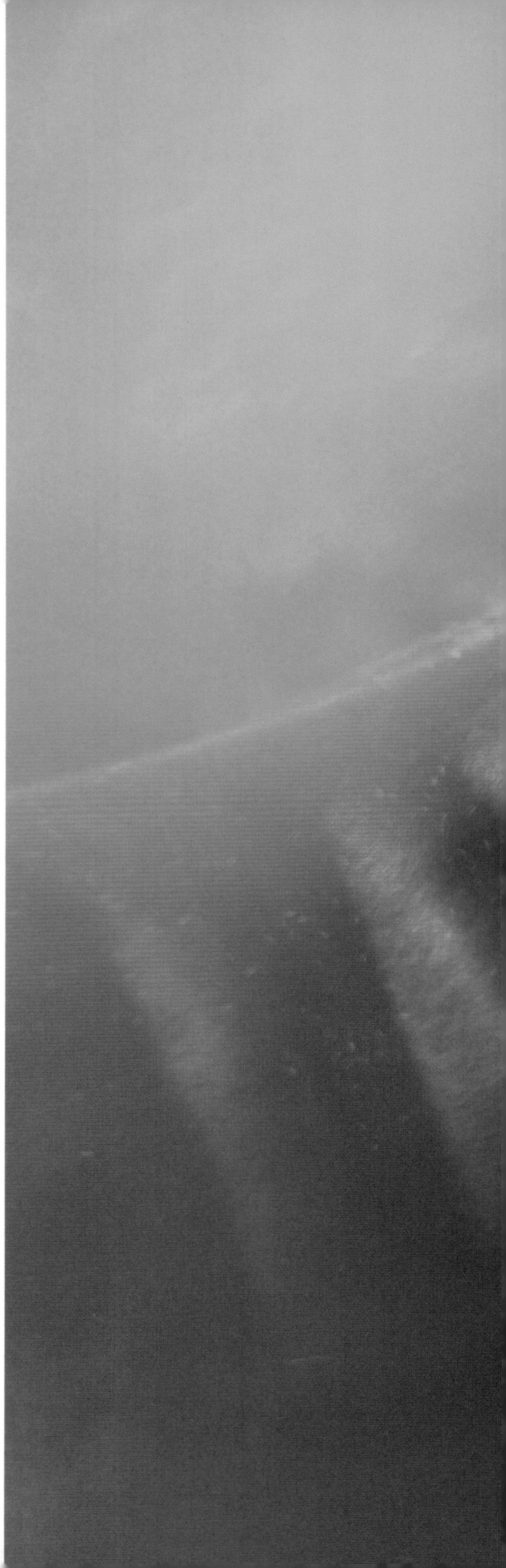

BODIES

WITH ONLY A FEW BRIEF REPORTS, unreliable drawings, and no specimen material to work from, the Swedish botanist Karl Linnaeus originally placed giraffes within the larger group of animals that included deer, sheep, and goats. In 1735, he named the species *Cervus camelopardalis*. Then the bodies began to arrive. Bodies: skins and skeletons that became the raw material of museum displays and scientific specimen collections, carefully preserved in the drawers, cabinets, and closets of a few zoological research institutions in England and continental Europe.

They came, first, from southern Africa, soon after a group of men led by Dutch explorers Hendrik Hop and Carel Brink rode their horses north of the Cape settlement into a territory sparsely inhabited by the Nama people. Europeans called the region Namaqualand. Brink kept a journal (later abridged to remove any reference to the unfortunate murder of one expedition member by another) that tells of sighting giraffes not long after the expedition camped at "Warmbath" (today's Warmbad) just north of the Orange River in modern Namibia. The date was October 5, 1761.

"Here the first giraffes were observed by us," notes Brink, adding that "one of these, a female, was killed and the young one with it captured." The dead female's lower body resembled that of an ordinary milk cow, he thought, with the head and neck looking like those of a horse. The men measured the body from head to hoof: 17 feet (about 5 meters). Comparing height to length of body produced a ratio of two and a half to one, and that remarkable proportion, along with the live animal's "extraordinary movements," made it altogether "unthinkable that this animal can be employed for any useful purpose." The giraffe was white around the neck and front, with "light brown checkered patches and diamond-shaped though more darkly brown patches on its hinder parts." She had two projecting horns covered with skin and hair; and it seemed as if the animal had been feeding high in the tall "red-wood trees," and "on account of the tall forepart of her body," she would have to crouch on her knees before she could graze on the ground.

The live youngster died a few days later, and they pulled off the skin to save as a specimen. Then, on October 16, they shot another giraffe—this time an adult male—and again measured the body. And in an entry dated November 22, the journal describes vast herds of grazing animals—not only giraffes, but also rhinoceroses, buffalos, and zebras, along with gemsboks, hartebeests, kudus, and wildebeests.[1]

The skin from Hop and Brink's expedition was sent, along with a sketch probably done by Brink, to zoologist J. N. S. Allamand at Leiden University in Holland. The skin and sketch were then used to model a more coherent artistic re-creation, the first of a southern giraffe, that appeared in the 1770 Dutch edition of Buffon's *Histoire naturelle*. George-Louis Leclerc, Comte de Buffon, was the distinguished director of the famed Jardin des Plantes in Paris, and his thirty-six-volume *Histoire naturelle* ranked among the best-known philosophical and scientific works of its day.

Following Hop and Brink came Gordon—Colonel Robert Jacob Gordon—who, during the two decades between 1770 and 1790, walked and rode his horse across a good deal of land in southern Africa, exploring the middle portion of the Orange River and, like Hop and Brink, moving north into Namaqualand.

Having originally served in the Scots Brigade for the Netherlands, Gordon came to Africa as a garrison soldier for the Dutch East India Company and commanded the Cape garrison between 1780 and 1795. He was also a dedicated naturalist and first-rate linguist (fluent in French, Dutch, English, Khoekhoe, and Xhosa), and he liked to hunt and to paint in watercolors. The latter set of interests provided more scientific material for the record. Gordon corresponded regularly with Professor Allamand in Leiden and was the first person to dissect a giraffe, describing for scientists the animal's anatomical interior. He also was first to send a giraffe skeleton overseas: a crateful of bones reassembled and recovered with skin to re-create a passable likeness of the original beast for the collection of Prince William V of Orange. The colonel sent four more skins to Allamand and also provided some extensive natural history notes to him and his colleague Arnaut Vosmaer. Vosmaer, in turn, is credited with having published, in 1786, the first scientific paper on giraffes, based significantly on the materials provided by Gordon.

In 1777, meanwhile, a twenty-one-year-old Scotsman named William Paterson shipped into the Cape Colony. Sponsored by Mary Eleanor Bowes, Countess of Strathmore, Paterson had come to find interesting or useful botanical specimens. He returned to England three years later not only with a major collection of plants, but also with enough written material for a book of his explorations and adventures, and enough skin and bones from a giraffe to justify their donation to the Royal College of Surgeons in London.

Immediately after the departure of the plant-gathering Scotsman there arrived a bird-collecting Frenchman. François Le Vaillant shipped out to the Cape in 1781, having been hired to shoot birds by the Dutch East India Company's treasurer, Jacob Temminck. During three expeditions between 1781 and 1784, the French collector acquired some two thousand bird skins for his wealthy Dutch patron: a remarkable assemblage that would become, in 1820 (when Temminck's son Coenraad became museum director), a centerpiece for the ornithological collection of the Dutch National Museum of Natural History in Leiden. Since so many of Le Vaillant's bird specimens were entirely new to science, he is perhaps best remembered for those feathery treasures, still commonly identified with names like Levaillant's Barbet, Levaillant's Cisticola, Levaillant's Cuckoo, Levaillant's Parrot, Levaillant's Tchagra, and Levaillant's Woodpecker.

But Le Vaillant is also remembered for his giraffe collecting, which he described in a book published in Paris in 1790, with a number of translations appearing in the next few years. Le Vaillant left Cape Town on that second expedition with around

Who would believe that such a conquest should excite transports bordering on madness? Troubles, fatigues, pressing wants, uncertainty of the future, and sometimes disgust of the past, disappeared altogether: all fled at the sight of this new prize. I could not satisfy my eyes with contemplating it. I measured its immense height. My eyes turned with astonishment from the animal destroyed to the instrument of destruction. I called and called my people one after another: and though any of them could have done as much, though we had killed animals of greater bulk, and much more dangerous, I was the first to kill this; with this I was about to enrich natural history; I was about to destroy romance, and establish a truth in my turn. **—FRANÇOIS LE VAILLANT, 1796**

twenty hired help (a few whites, the majority indigenous "Hottentot," or Khoekhoe), three horses, thirteen dogs, eleven goats, three milk cows, and thirty-six oxen for pulling the wagons that carried most of the supplies. A pet baboon named Kees also came along "for my amusement," as Le Vaillant would write, "and, I may also say, for society."[2]

Like others before him, Le Vaillant traveled to Namaqualand, that great open region to the northwest, and not long after crossing the Orange River, he saw a giraffe. "Ravished with joy" upon catching sight of the creature, Le Vaillant leaped onto his horse and, accompanied by an African assistant and a pack of yapping dogs, set off in heated pursuit. The animal "trotted on lightly without exerting himself in the least, while we galloped after, firing occasionally at him; however, he insensibly gained upon us, so that after a chase of three hours, our horses being completely out of breath, we were obliged to stop, and soon lost sight of him."[3]

They saw five more giraffes the next day, but the peculiar creatures "employed so many wiles, that, after we had hunted them the whole day, they escaped us through the favour of the night." Le Vaillant was "grieved at this bad success," but he was even more concerned about the expedition's food stores, which were now down to a few pounds of rotting hippopotamus meat—insufficient for a party that by then included twenty-six men. The next day, therefore, the Frenchman set out at sunrise, accompanied by a single assistant and intending to find some game to feed the expedition. After riding and walking their horses for several hours, the two men rounded a hill to discover seven giraffes. The dogs instantly rushed out, while Le Vaillant and his associate pounded after them on the horses.

Six of the animals quickly fled, but the seventh was soon "surrounded by the dogs, and endeavoring by forcible kicks to drive them off. I had only the trouble to alight, and brought him to the ground with a single shot." Having managed so directly to kill the animal, Le Vaillant was beside himself with excitement and self-congratulation. "Who would believe that such a conquest should excite transports in my mind bordering on madness?" he would recollect. "Trouble, fatigues, pressing wants, uncertainty of the future, and sometimes disgust of the past, disappeared altogether: all fled at the sight of this new prize." The hunter paused to measure the stretched-out carcass in all its surprising immensity; and, finally, he took the measure of his own grand accomplishment in all its turgid glory: "I was the first to kill this; with this I was about to enrich natural history; I was about to destroy romance, and establish a truth in my turn."[4]

Of course, Le Vaillant was not actually the first person to kill a giraffe on behalf of science. But he was among the first. And he may well have been the first to take his task so seriously, so carefully measuring and sketching the carcass, so methodically preserving the skull and hooves and bones and skin, all with an eye to their later utility in the great European halls of learning—even as the rest of the expedition, having gone hungry for the last day and a half and now positively ravenous, impatiently sharpened their knives.

—

Careful measurements and good specimen materials became the first toolkits of zoological anatomists, and in the nineteenth century, more hunters—including, now, battalions of puerile European sportsmen, gone out to the colonies for a bit of sweat and blood—added more specimen materials. Meanwhile, the gradual development of zoo populations brought in yet more anatomical opportunities, including the potential for close observation of living animals and, of course, for clean dissections of deceased ones.

Pasha Muhammad Ali of Egypt may be credited with originating the European zoo population, having sent those early zarafas to King Charles X in France and King George IV in England. In fact, the pasha soon sent a third giraffe to Vienna, where she was put on exhibit at the Schönbrunn Zoo. Although the one in England died within a year of her arrival, she was replaced in 1830 by a pair of giraffes; and, as I noted in the previous chapter, the French zarafa was given a companion in 1839. By the middle of the century, European zoos had enough giraffes to breed them successfully.[5]

The measurement of giraffe specimens ordinarily began with their height. Given that giraffes are the tallest animals on earth, it is natural to wonder about an individual's height, and—at least for those who hunted for personal pleasure and self-aggrandizement—to look for great trophies and world records. The record is 19.29 feet (5.88 meters), toes to top.[6] A more ordinary height would be around 14.8 feet (4.5 meters), which is one estimate for the typical adult male in Nairobi National Park in Kenya. But an adult giraffe can readily extend his or her feeding reach by an additional 3.6 feet (1.1 meters) or so simply by elevating the snout and stretching out a long, prehensile tongue.[7]

The world's tallest animal acquires that unique height through long legs, a long neck, and a high and massive chest, and those characteristics in turn mean a very large body. Aside from being uniquely tall, then, a giraffe is also very heavy. With adult males weighing around 2,650 pounds (1,200 kilograms), among terrestrial animals they are second in mass only to elephants.[8]

—

But why the great height? And why the long neck? After the publication of Charles Darwin's *On the Origin of Species* in 1859, it became reasonable to ask such questions. Eventually, it even became possible to answer them in a satisfactory if still tentative fashion.

We now understand that contemporary giraffes—*Giraffa camelopardalis*—are the result of an evolutionary process that happened gradually and over a long time. It also involved the simple yet steady modification of already existing structures. Consider, for example, the legs and feet. These are essentially, with a few important modifications, the same legs and feet found among other mammals, with the most obvious distinction being size. Or consider the neck. The seven cervical vertebrae that make up the bony structure of your neck and mine, and the neck of virtually every other mammal except for manatees and sloths, will be found in a giraffe's neck. Same number of vertebrae. Same shape and function, more or less—with the "less" part including the fact that the first adjacent vertebra supporting the ribs and thorax (that is, the first thoracic vertebra) has turned to connect with and support the neck in a way that the final cervical vertebra typically does for other mammals.[9] But giraffe neck bones are lengthened proportionately, compared to similar-sized animals, to a substantial degree. That lengthening may be related to a simple, genetically-programmed increase in responsiveness to normal growth hormones.[10]

Once we think of tallness and long-neckedness as evolutionary gifts, we can then intelligently wonder why they came about. What were the selective advantages of these two related developments?

The ordinary workings of mutation and genetic shuffling provide variety in any group of animals. Some individuals in a group will be bigger or taller or longer-necked than others. In the case of giraffes, the evolutionary process producing unprecedented tallness and long-neckedness happened because those individuals who were slightly taller and had slightly longer necks than others were, generation after generation, at least slightly more likely to succeed reproductively. They had, on average, more surviving and reproducing offspring than other giraffes who were comparatively less tall and less long-necked. Even slightly more reproducing offspring, on average, would mean that any genes associated with tallness and an elongated neck would be more fully transmitted into subsequent generations. If this process continues consistently, then the species as a whole will gradually become taller and longer-necked. So what were the advantages acquired by those tall and long-necked individuals that selected them to become the more successful reproducers?

Increased predator avoidance. Giraffes are big enough to fight off most predators, with the exception of lions.[11] Their great size inhibits speed (through inertia and the comparatively slow contraction of larger muscles), but their longer legs promote it; and giraffes can gallop in bursts of up to 35 miles (56 kilometers) an hour.[12] At the same time, their greater height improves the capacity to see predators from a distance. Indeed, the giraffe's watchtower physique seems especially suited to gently rolling savannas, where predators—lions, for example—are often able to approach their prey by moving quietly behind a small rise or minor hill. Giraffes can often see that predator on the other side of that hill, which gives them a distinct advantage over individuals of other species, while taller giraffes have the same kind of advantage over shorter ones.

Improved thermoregulation. The upward stretch provides a second advantage in terms of heat dissipation. Compared to other large African mammals—hippos, rhinos, elephants—a giraffe's body is less rounded, which exposes a greater surface area per unit of volume; and it is at the surface that the body heat of a warm-blooded animal living in a hot climate will dissipate. In addition, a giraffe's narrower distribution up, rather than out, minimizes the surface area exposed to an overhead sun. These shape advantages may seem minor or theoretical, but they could be important. It is interesting to note that while hippos routinely soak themselves in water, and while rhinos and elephants wallow in mud or seek out shade, giraffes seem not as dependent on water or mud or shade, and they have a great tolerance for high heat and direct sun. In a study where zoo giraffes had a clear choice of sun or shade, at the Taronga Zoo in Sydney, Australia, giraffes chose the shade only 12 percent of the time.[13]

Expanded feeding opportunity. The feeding competition theory was first suggested by Darwin, and the concept is simple enough. Giraffes, by becoming tall and evolving the long neck, opened up a better and more reliable food source not available to other browsers, giving a competitive advantage to taller over shorter. This advantage seems clear today, since modern giraffes, while they may lower their necks and comfortably browse at only 1.6 feet or so (around 0.5 meters) off the ground, will also feed at close to 15 feet (4.5 meters) above the ground. Their next tallest browsing competitors—kudus—can reach for food only as high as 6.5 feet (2 meters) above the ground.[14] The theory, however, describes a contemporary situation rather than that of an extended evolutionary past, when ancestral giraffes were actually competing with a number of other high-level browsers.[15] Still, a tall animal can choose high or low, whereas the less tall can only choose low, and that choice advantage might be decisive during times of drought when food is in short supply. It is true that such a drought would eliminate from a population most young and many adult females, who are smaller and shorter than adult males, but only a few females are required in order for the larger population to recover and even flourish after a drought has ended. And even though mortality among the young could be very high in such times, most surviving adults have many years beyond the span of a typical drought in which to reproduce and raise additional offspring. The tall survivors of a drought, then, could very well pass some of their gene-based inclination to tallness on to the next generation.

Greater reproductive opportunity for males. As with most mammals, giraffe males compete with one another for sexual access to fertile females, and the adult males spend a good deal of their time following females and testing them, through smell and taste, to determine the approach of maximum fertility. Once a female near maximal fertility has been discovered, the males will be inclined to compete; and, of course, the winner of the competition will become not only a sexual and reproductive winner but also an evolutionary one. The male who succeeds more often in this competition, in other words, will contribute more to the gene pool of the next generation. And it happens that giraffe males fight each other by using their necks and heads as weapons, a behavior that is sometimes generally described as *necking*.

Most of the time it seems gentle, even affectionate, as if the males are playing or lightly testing each other or gently sparring. On occasion, however, the necking sessions become more serious, and in cases of open competition over a fertile female, they can turn very serious indeed. A male giraffe's neck can be more than 7 feet (2 meters) long.[16] The neck combined with the head weighs around 550 pounds (250 kilograms).[17] The neck is unusually strong and flexible, with the seven vertebrae connected to each other through a tight ball-and-socket-style articulation that, though common among reptiles, is unusual among mammals.[18] The head is based on a very large and well-armored (especially among males) skull, with two horns and often other areas of thickening and projection. An old male's skull may weigh up to 30 pounds (13 kilograms), while an adult female's could weigh around a third that much.[19] Females do not compete with each other through necking battles, and yet neither the skulls of males or females are as heavy as they might be. Nature has lightened those large skulls with unusually large air pockets, or sinuses, that increase volume without adding mass.[20]

What drove the evolution of that great height and long neck? All four theories seem more or less reasonable, although the first three fail to explain why only certain ancestral giraffe species began to develop the height and neck so characteristic of contemporary giraffes. The fourth theory, the one based on sexual competition among males, seems more robust, if only because it considers a behavior unique to giraffes.

On the other hand, the sexual-competition-among-males theory fails to explain one important thing: Why have giraffe females also retained the long neck? Since the females never fight each other or males or drive off potential predators with their necks or heads, why should they have acquired and retained necks any more substantial than, say, those of zebras?[21] The answer to such questions and conundrums requires us to appreciate complexity. We should not think of these four advantages, and others, in isolation. It is likely that a number of complex advantages drove the evolution of long necks and great height at one time or another in the giraffe's evolutionary past, and they would have done so in complex combinations and to various degrees.

—

The reason other species occupying a similar ecological niche did not also discover the wonderful advantages of long necks and great height is closely related to the reason giraffes have not continued to grow increasingly longer and taller. The same advantageously upward trends in body size and shape will also create a number of serious disadvantages requiring serious compensatory developments. Nothing in life is free.

For example, that great neck adds tremendous weight, which requires an appropriately massive structure to support it. Meanwhile, moving from a standing to a sitting or lying position becomes complicated (folding the forelegs first, then the hind ones), while rising from sitting or lying requires an added shifting of weight, the neck now being used as a counterbalance (vigorously lunging forward to help straighten and raise the rear legs, then lunging back to help with the front legs). And because the legs and torso are longer than the neck, the creature must awkwardly bend or bow those legs to engage in the simple act of drinking water from a pool or stream.

When a giraffe gallops (front legs moving inside hind legs in a controlled series of leaps), that massive neck and head shift back and forth like a gracefully undulating pendulum, adding impetus to the forward motion at the start of each leap, drawing back as the feet touch down. Meanwhile, the lengthened legs and shortened trunk require an unusual gait that probably serves to keep the front and rear legs from striking each other. In walking, a giraffe moves both legs on a side almost simultaneously, with the front foot rising from the ground just as the same-sided rear foot begins to swing forward.[22] This lateral gait contrasts with that of most other quadrupeds, who tend to walk diagonally (front foot on one side moving in near synchronicity with the rear foot on the other side).[23] The gait affects weight distribution. Any animal, such as a deer, who distributes weight more fully on diagonal legs acquires better balance and a greater capacity to change direction quickly than the animal, such as a giraffe, who distributes weight laterally.[24]

Bigger also requires more nutrition, which in turn requires a giraffe to spend more time searching for and consuming food, day and night—with the average time set aside for sleep being less than two hours a day.[25] Bigger also means, for a giraffe at least, longer legs, with a high knee and a disproportionate elongation of the foot and calf bones; and because those long bones acquire their strength through increased density rather than thickness, giraffes require more calcium than other herbivores. During the first four to five years of life, as a male giraffe moves from infancy to full adulthood, his total amount of skeletal calcium increases tenfold, requiring an average calcium absorption of around twenty grams per day, or around four times the amount a growing human needs. This in turn forces the giraffe to depend on high-calcium legumes such as acacia trees and shrubs.[26]

But perhaps the most obvious disadvantage of great height is the circulatory one. It must be fine indeed to have such a high vantage point from which to look out over the world, but hoisting the brain up around 8.2 feet (2.5 or so meters) above the heart necessitates a very strong heart to keep the blood pressure high. A giraffe's heart, weighing some 25 pounds (around 11.5 kilograms), is sufficient for the task, but there are other problems. If we imagine a giraffe's circulatory system as a tall column of blood, we can appreciate how the weight of that blood down at the lower legs and near the feet could burst through the walls of an ordinary vein or artery. For a giraffe, evolution has thickened the vein and artery walls and wrapped the lower limbs in a reinforcing sheath of smooth muscle.[27]

Then there is the problem of rapid and extreme shifts in blood pressure, as, for instance, when a giraffe suddenly lowers her head from a high feeding position to a low drinking one—whereupon a large amount of blood is suddenly impelled to rush down toward the brain. Humans experience a comparable change in blood pressure when strapped in a fighter jet gone into a quick dive, where blood rushing into the brain will produce blackouts. But all the large veins in a giraffe's circulatory system include valves that prevent the blood from flowing backward when the head is suddenly lowered; and within the giraffe's neck, the major veins and arteries are diverted into a complex network—known as the rete mirabile (miraculous net)—that, operating like a large, dynamic sponge, inhibits the flow of blood into the brain when the head is suddenly lowered, and inhibits flow away from the brain when the head is suddenly raised.[28]

The image that opens this chapter shows a Masai giraffe feeding on vegetation that no other earthbound mammal except an elephant can reach. Note that the giraffe has extended his reach significantly with a lift of the head; the long tongue can provide further extension. The unnaturally flattened underside of the tree indicates that giraffes have thoroughly fed here at their maximum reach.

The images below continue a meditation on the advantages and disadvantages of that great height and reach. As the first in the series suggests, a giraffe can commune with the birds—in this case, a secretary bird (*Sagittarius serpentarius*). The subsequent picture (three Masai giraffes keeping track of a suspiciously approaching human photographer) shows how, more pragmatically, the watchtower physique provides a prime advantage in predator avoidance. One major disadvantage of that same physique is the awkward problem of getting a drink of water, as demonstrated by the next image.

The next two photographs feature other giraffe extremities: long- and dense-boned legs and a long, prehensile tongue. The final two images illustrate the giraffe walk (legs alternating side to side, as shown by the middle animal in the group) and gallop (legs alternating front to back).

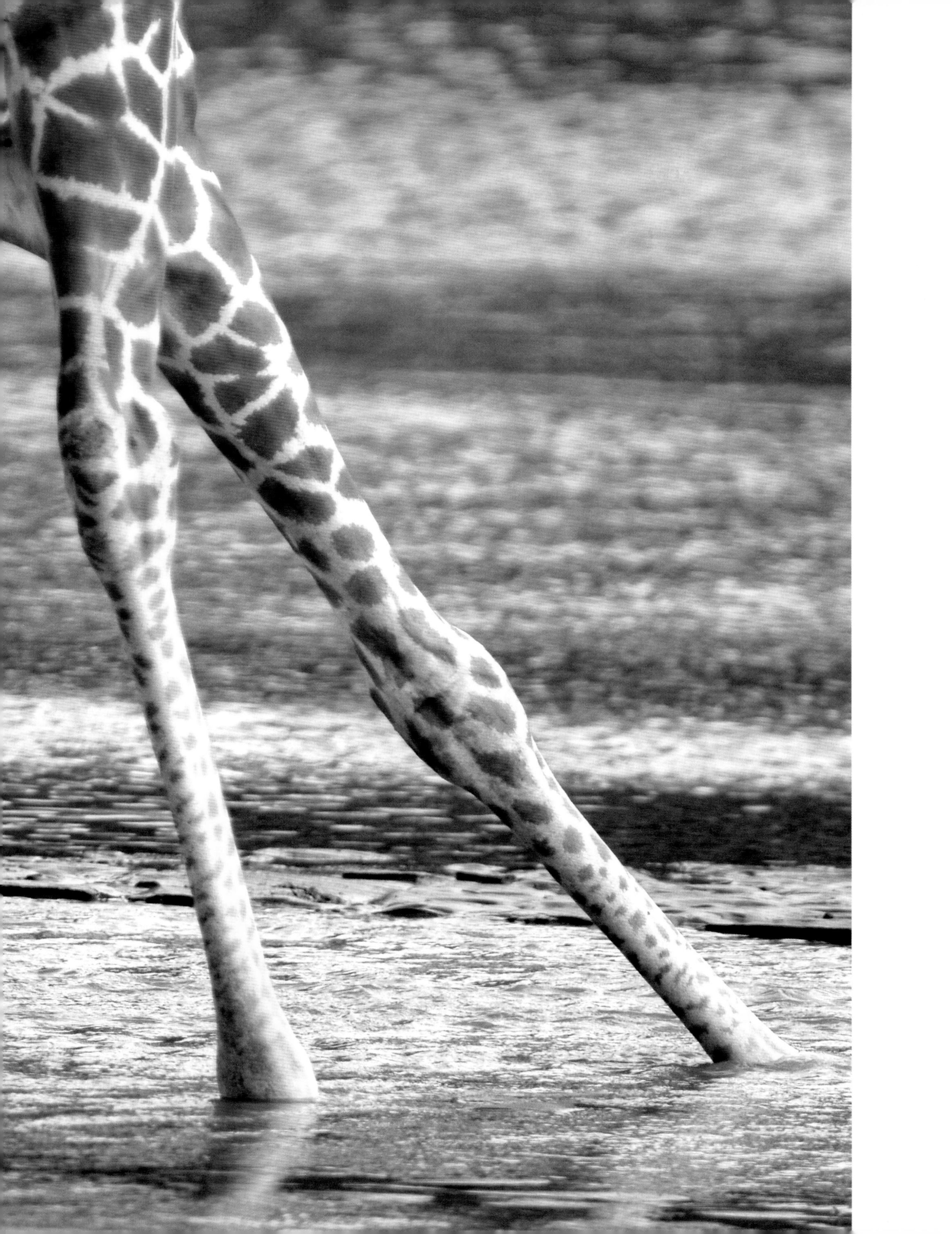

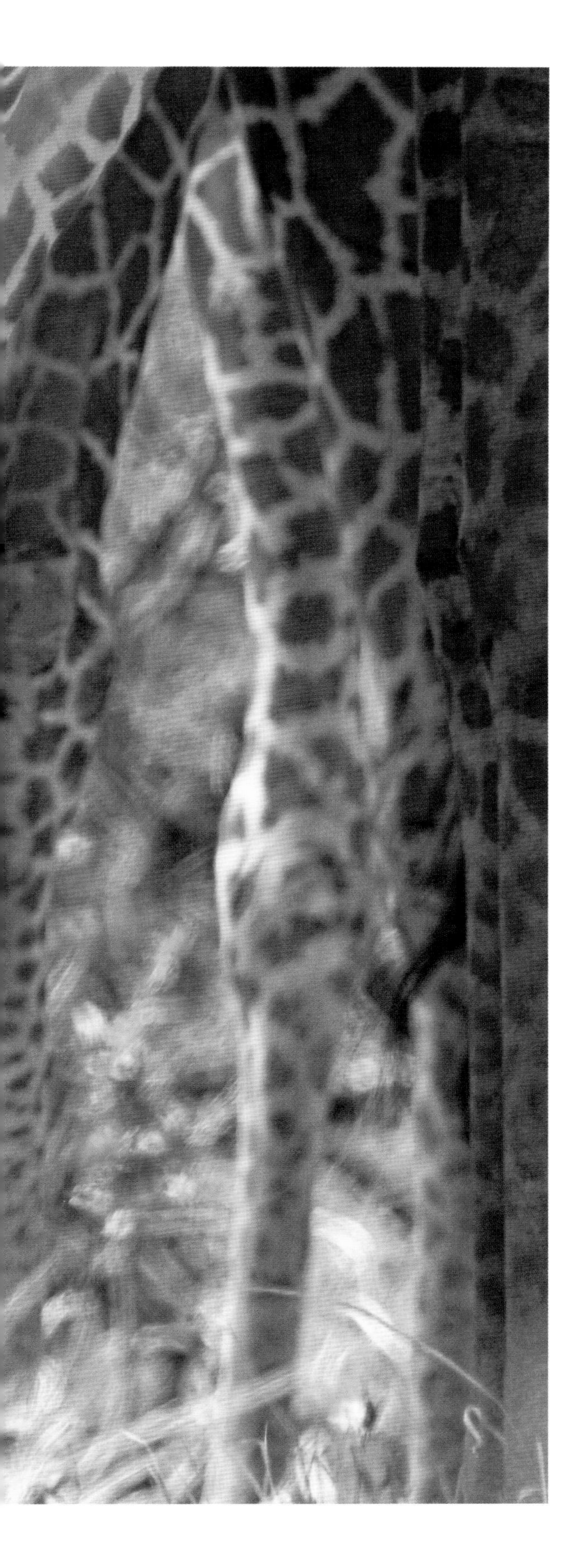

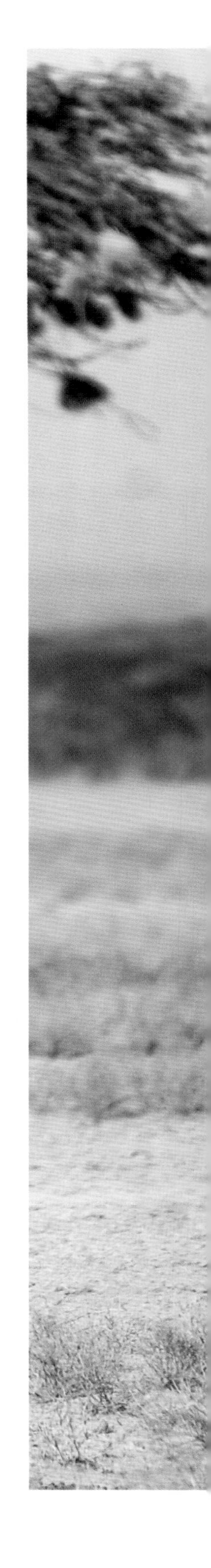

BEHAVIORS

IT COULD HAVE BEEN THE GREAT HEIGHT. Maybe it was the huge, solemn face or the intelligent gaze from a dark and glistening orb. It might have been a strangely swaying gentleness or yet some other aspect of the towering drama: a sudden burst of movement, for instance, a rapid shift in pattern or a clattering clash of light and dark. To a small child everyone is a giant, certainly, but for this particular child to find the giant among giants must have been astonishing, and that this particular giant among giants seemed a benevolent one would have been deeply reassuring as well. Whatever the origins of her unusual attraction, the two-year-old girl fundamentally understood, as she would later recall, that "all the other treasures of the Brookfield Zoo were nothing compared to the wonder of the giraffe."[1]

Anne Innis grew up in Toronto, Canada, the youngest of four children. Her father was a professor of economic history and graduate dean at the University of Toronto. Her mother was the author of a novel and short stories as well as nonfiction books that included several Canadian histories. There were no giraffes in Canada at the time, though, and so the child first saw a giraffe when, during a visit to the United States, she was taken to Chicago's Brookfield Zoo. Her infantile attraction turned into a childish enthusiasm, and it remained as such until, growing with the child, it emerged as a life passion.

She entered the University of Toronto in 1951, enrolling in a science program that she hoped would eventually—after the requisite language studies, along with courses in botany, chemistry, geology, math, physics—lead her to zoology and, thus, to giraffes. Zoology, however, required the memorization of Latinized names; the gathering of botanical collections; the detailed study and critical examination of amoebas, worms, frogs, rabbits. Never giraffes. And never living, breathing, *behaving* animals.

Study live animals? It was possible. Field work was still a rare oddity, even though, by the mid-1950s, a few scientists had successfully studied the behaviors of a few accessible species of insects, fish, and birds in Europe. A few others had done research on some monkeys and a couple of other kinds of animals in Japan and Central America. But before Jane Goodall, Dian Fossey, George Schaller, or any of a half dozen other well-known pioneers arrived on the continent, no Westerner had even tried to initiate an extended behavioral study of African wild animals. Anne Innis would be the first. Of course, being first meant that when she graduated at the age of twenty-two from the University of Toronto—still determined to study giraffes—there was no precedent, no path to follow. There was no program. No mentoring professor. No financing. No infrastructure. And no particular sympathy for a young woman trying to do what even a well-established male scientist might only wildly consider.

She decided to finance the research herself. She had saved money in her undergraduate years by feeding mice, cleaning bones, recording the measurements of fox skulls at the Royal Ontario Museum, and running laboratory demonstrations at the University of Toronto. Her savings, combined with a small gift from her mother, would get her to Africa, but once there, how to find the giraffes? And where might she find a place to stay, financial and logistical support while she was there, and perhaps a sympathetic colleague or two? Her applications to the nonexistent position of First Serious Field Scientist in Africa consisted of a series of exploratory letters sent to a few likely places, mainly game departments in Uganda, Kenya, and Tanganyika. But with no positive response to the letters, she moved on to graduate studies at Toronto, working for her master's degree on a problem in mouse genetics, while noting ironically that "my friends had a wonderful time joking about research done on mice as a prelude to studying giraffe."[2]

Finally she received a welcoming letter from a generous zoologist at Rhodes University in South Africa. Professor Jakes Ewer wrote that he knew of a farmer, or farm manager, who had seen giraffes roaming over parts of his 20,000-acre cattle ranch and citrus farm, called Fleur de Lys, which was located a thousand miles (about 1,600 kilometers) northeast of the university in the Transvaal Province, not far from Kruger National Park. The ranch was sectioned with fencing, but the giraffes—around a hundred of them—had learned to step or jump over the fences, so for the most part they roamed freely across the land, while at the same time the farmer had set aside one large piece of his property to serve as a protected game reserve.[3] Professor Ewer, in fact, had already taken the liberty of corresponding with the farm manager, Alexander Matthew, who had declared that he would welcome a visiting zoologist. The only concern, as Ewer went on to explain in his letter to Innis, was that he had insightfully neglected to mention that the zoologist in question was a young woman, and so Mr. Matthew—married to a temporarily absent wife—had invited A. C. Innis to stay at his remote ranch and citrus farm based on the mistaken presumption that she was a he.

Innis sailed from Montreal to London in May of 1956. Then, soon before boarding a Union Castle ship bound for Africa, she wrote to the farmer directly, reassuring him that she was a serious and professionally trained zoologist while explaining that she was actually a woman and not, as he must have imagined, a man. She hoped he would accept that additional information positively, but in any case, by the time he received her letter, she would already be in South Africa, and he could respond in care of Professor Ewer at Rhodes University.

Once settled in as a houseguest of the Ewers at Rhodes, Innis paid £200 for a beat-up green 1950 Ford Prefect, which she affectionately named Camelo, a shortened version of the species name for giraffes (*Giraffa camelopardalis*). And then, after receiving a confirming invitation from the rancher, she loaded her few possessions into the car and began the thousand-mile drive north and east to Fleur de Lys. The young zoologist arrived late at night on the third day, her car having broken down on a rough dirt road about five miles (eight kilometers) short of the destination. Appropriately concerned about the dangers of walking alone on a moonless night—poisonous snakes, lions, hyenas, and the like—she nevertheless placed her pajamas in a backpack, her wallet in a pocket, and set out on foot.

The next morning, she joined Matthews and one of the citrus orchard managers for a welcoming breakfast of fish, toast, cold cereal, and papaya. In exchange for secretarial work, Matthews had offered her room and board; he also hoped she would be able to provide him with a bit of botanical information about the local grasses and zoological information about the local game animals. Now he handed her a pair of binoculars, offered her the use of a 16-millimeter camera and film, and then, politely, he asked her to clarify her intentions: "What are your plans?"

She responded simply: "To study the giraffe."

He pressed: "What aspect of the giraffe?"

"Everything I can think of," she replied, adding, "All aspects."

"How long will it take?"

"Well, that depends on how it works out, I guess."

—

A couple of months after Innis arrived at Fleur de Lys, a government official shot a giraffe who, so it was said, had been inconveniently blocking a hardly used road. The shooting was unfortunate, but it gave the young zoologist her first opportunity to dissect a giraffe—with the help of fourteen men working for three hours to divide the carcass into manageable pieces and tote it back to the ranch. The workers sang as they hacked and tugged, perhaps happily anticipating the large pieces of meat they would be taking home to their families, meanwhile turning that section of the road into something resembling "a butcher's shop, with men staggering about carrying bloody segments of leg, or liver, or abdomen."[4]

Workers cut and yanked away the skin, piecemeal, and placed it all in the truck. The neck was chopped like firewood into seven equal pieces, one for each vertebra, "each surrounded with its casing of muscles, nerves, and ligaments," then loaded in the truck. Six workers were needed to lift the stomach into a basin. Two men lugged the animal's head. The heart was placed in a large bucket. The intestines were coiled around the bucket. And so on. Back at the ranch, after the workers had finished unloading their disassembled giraffe and bicycled away, each with his own bundle of meat tucked under a free arm, the zoologist went to work. She identified three tick species from the giraffe's skin, preserving them in jars of cheap whiskey. She cut open the stomach, hoping to find identifiable remnants of food, instead discovering—in the first chamber, the rumen—"about a hundred pounds of small twigs and 'guck.'" The contents of the remaining three chambers of the stomach were less distinctive, an increasingly refined progression of chlorophyll-tinged fluids. Innis uncoiled the intestine, then photographed and measured it with around eighty flips of an old yardstick. The heart, she concluded, was approximately the length of her forearm and hand combined. The aorta was "much bigger in diameter than a silver dollar," the blood "sloshing around by the gallon."[5]

Another time, using the camera and film provided by Matthew, Innis made a series of movies—of giraffes walking, galloping, lying down, standing up, drinking, eating, ruminating, and so on—and thus she gathered an early film record of giraffe movements while providing the opportunity for a later frame-by-frame analysis of the biomechanics involved in them. The unusual lateral gait of a walking giraffe has been written about since classical times, but no one had studied that walking gait—or the galloping sequence—with careful precision until Innis reviewed her film. Nor had anyone before Innis assessed the biomechanics of giraffe walking and running as it was integrated with the shifting-pendulum effect of a great neck and heavy head.

Of course, zoologists had dissected giraffes before. They had plenty of opportunities to do so, in American and European zoos, for example, under far better conditions and using far more precise instruments. And of course it was possible for almost anyone to film giraffes—zoo giraffes—in motion as a reasonable approach to analyzing biomechanics. Innis understood that. While she had been a guest associated with Rhodes University, Innis had consulted the university's zoological library to survey all the scientific literature she could find on giraffes, confirming that it consisted almost entirely of anatomical and physiological observations based largely on dead animals or zoo exhibits. "There are," as she later phrased it, "lengthy discussions on the presence or absence of a gall bladder, on the exact arrangement of blood vessels connected with the aortic arch, and on the unusually wide structure of the lower canine tooth, which enables a paleontologist to recognize a member of the giraffe family from that tooth alone. Odd, I think. When I visualize giraffe, I think of them as living ani-

I capture [on film] Cream browsing at a tree, his head filling the entire frame with his long black tongue reaching up, straining to curl around and grasp a leaf above. When he moves over to a thorn bush, I film him chewing while using both his tongue and his mobile lips, covered with dense fur to protect them from the thorns, to separate leaves on a twig from the thorns. Sometimes he closes his mouth over an entire twig and jerks his head away, stripping the leaves into his mouth. This combing action is aided by the unique extra lobe of the outer tooth of the lower incisor row, which increases the row's width; these incisor teeth meet the upper bony plate which, during evolution, has replaced the upper incisors in most large herbivores. As I'd noticed earlier, the giraffe takes only a few bites from one plant before moving on to the next. **—ANNE INNIS DAGG, 2006**

mals wandering freely on the African savannahs, not as soulless assortments of unusual organs, blood vessels, and teeth."[6]

Living animals wandering freely. While her experience in dissecting a recently killed giraffe was interesting, then, and while her film study was useful, the unique significance of Anne Innis's contribution is categorically different: She was the first person to study living giraffes wandering freely. She did so by considering carefully, methodically, quantifiably—scientifically—their behavior.

—

Her methods were simple enough. First she learned to identify the few obvious differences among them, such as those related to age and sex. Estimating age amounted to a series of increasingly sophisticated guesses based largely on relative size, with newborns about six feet tall. Sex? From a distance, often the quickest way to distinguish adult males from females was to consider the horns. All giraffes are born with a wavy crown of black hair on the tops of their horns, which the adult males wear off in the process of banging their heads against each other. Their horns are always bare at the tops, while adult females retain that lovely crown of wavy hair.

Then Innis began learning to recognize specific individuals, intuitively using whatever characteristics were the most obvious. Having seen two adults browsing in an area nearby, eating away at some acacia leaves, she noticed first that the female had "jaunty" crowns of hair on top of her horns. Innis called her Pom-Pom. The male beside Pom-Pom had spots that looked like a child's drawing of stars. Let him be Star. A male with distinctive lumps on his lower neck became Lumpy, while another male who had walked erratically as the result of a snare injury was Limpy. A fourth male, with body spots on a darker cream instead of the usual white background, became Cream. So it went. Later researchers would learn to base individual identification on a giraffe's pelage pattern—the spot design—which happens to be as unique and unchanging as a person's fingerprint. These can be recorded photographically and carried around in an identification album.

Innis worked with quickly sketched position maps, making an *X* for any male and an *O* for any female, then modifying those symbols as appropriate with a *y*, indicating young, and a *b* for baby. That was a rough way of recording the positions of and, possibly, social relationships among individuals. Innis elaborated on the data by noting, once every five minutes, each individual's behavior based on five categories of likely behaviors: eating, ruminating, lying down, walking about, and socially interacting.[7] It was a start, at least, although Innis would soon feel overwhelmed by the sheer volume of data she was collecting. A more serious problem with her early system was that the categories were potentially overlapping (a giraffe can lie down and ruminate at the same time) and limited (omitting, for example, vigilance behavior).

Being so big forces giraffes to spend much of their time eating, which Innis found "boring to watch." Nevertheless, she made a couple of interesting observations about their feeding style: that they seemed to feed selectively, moving quickly from tree to tree and bush to bush, and that they would often swipe or comb away leaves from a twig, making good use of that pair of incisor-like, double-lobed lower canines as part of the comb. She noted the sharp thorns found on several of the food species; and she began to gather information about what plant species these animals preferred.

Eating was, in fact, the most common activity, taking place at any hour of the day. As the sun rose in the morning, some three-fourths of her study subjects would be eating, while at dusk nine out of ten would likewise be chewing on their favored leaves and twigs. Around noon and into the hot early afternoon was a time of minimized eating, even though more than half the animals would still be engaged in it, and maximized ruminating.[8]

To identify the food species eaten at Fleur de Lys ultimately would require an expert botanist; for the moment, Innis simply watched until the giraffes were done feeding at a particular tree or bush. Once they had moved away, she moved in and tied a yellow strip of cloth to the plant, indicating it as a food plant. Later, when she had the time, she would gather fresh leaves from each of those marked plants and bring the collection to a botanist working at the National Herbarium in Pretoria for a full analysis. In this way she made a preliminary identification of the diet of giraffes at Fleur de Lys. It included some thirty-two different species of plants, which was a much wider variety, she would subsequently conclude, than during the wet season, when food was more abundant and the giraffes could concentrate on their favorites.[9] Later scientists would confirm that giraffes are adaptable enough to consume a quite large variety of species of plants—over a hundred, in some cases—eating mainly leaves and twigs but also, occasionally, flowers, fruits, and bark.[10]

—

Fleur de Lys was a less than ideal place to observe sexual behavior among giraffes, because for some reason there were many males and comparatively few females in the population. Innis observed mating once, although she also several times watched the urine-testing behavior that males characteristically do while searching for a fertile female.

Normally, the female giraffe cycles into her fertile period every two weeks.[11] During her days of maximum fertility she will allow or even perhaps encourage a male to mate with her. Not surprisingly, the males spend a good deal of time searching for fertile females, and their search largely consists of testing the chemical content of urine. The male will approach a female, lower his head to grasp her tail with his lips or to nuzzle her hindquarters, and thereby stimulate her to urinate. After collecting a sample of the urine in his mouth, the male throws back his head, and, curling back the upper lip in a peculiar way—employing the flehmen response—he considers the chemistry of the urine through his vomeronasal organ.

Flehmen is a word of German origin (used as noun, adjective, or verb), and it refers to that curled-back upper lip, which produces a distinctive grimace or smirk that happens to resemble the facial expression we humans make in response to a powerful odor. In giraffes—and many other species, including cats, horses, and elephants—the flehmen response involves a contraction of upper lip muscles that flares open a pair of small ducts located in the roof of the mouth. The opened ducts serve, in essence, as a tiny pair of secondary nostrils leading to a second, separate, highly specialized olfactory system—known as the vomeronasal organ—which is uniquely sensitive to the chemical signature of pheromones and leads directly to the parts of the brain associated with sex and aggression.

Having discovered that the female is fertile, then, the male approaches her from behind, and—presuming she is cooperative rather than shy or resistant—he rears up and presses his forelegs to her back, moving her into a more favorable position while steadying himself as he draws forward, then enters and thrusts briefly. The act is short and may not satisfy the expectations of a human observer. One author has suggested that giraffe sex must be "singularly dull."[12] Innis herself was certainly pleased by the observation yet disappointed by the act, wondering, as she would later recollect: "But where is the romance, the necking and tender rubbing of bodies that occurs in homosexual encounters? Did I miss these?"[13]

Homosexual encounters? The particular encounter she refers to here, which she witnessed a month or two earlier, completely surprised her; she may have been the first field zoologist to note homosexual behavior in any animal. It began with a sparring match. Giraffe males often spar with each other—that

It's another hot day, the temperature over ninety degrees Fahrenheit, and as usual I'm wilting in torrid Camelo, making notes every five minutes of what all the giraffe I can see in the Game Reserve are doing and allowing myself a sip of water from my tepid canteen immediately afterward. In the distance I spy Star, another male, and a female walking restlessly about. What are they up to? All the other giraffe are stationary and as drowsy as I feel. After circling her several times, Star stands directly behind the female, who has pulled aside her tail, raises his head and then mounts her, sliding his front legs quickly forward on either side of her flanks. A few seconds later she gives a short run forward, forcing him to dismount. The second male, engrossed with this enterprise, moves forward. They're mating, I think in excitement. But where is the romance, the necking and tender rubbing of bodies that occurs in homosexual encounters? Did I miss these?

—ANNE INNIS DAGG, 2006

is, lightly test one another's power and strength in a head-to-head, neck-against-neck mock battle. This necking behavior, as she called it, is not so different from the play-fighting that happens among a number of other species. Elephant males spar with their tusks and trunks. Human males lightly wrestle in the locker room. Giraffe males neck.[14]

As Innis noted, necking can appear gentle and friendly, and often, at least in the case of giraffes, done with a seemingly erotic quality that may lead to sexual contact. She observed the behavior among two males she was already familiar with, Star and Lumpy, whose necking, she recognized, turned gentle and light and very slow moving. It really looked affectionate, and it concluded with an erect penis and a brief mounting that left the human observer completely amazed and, since she was a young and comparatively innocent woman raised in an innocent culture, embarrassed. She had never in her life even spoken the word *homosexual*, and she would have been "embarrassed to death" to describe what she had seen to anyone at Fleur de Lys.[15] She included a description of the event in her first scientific report about the giraffes, published in 1958,[16] but she never mentioned it to anyone at the farm or—for two decades following that scientific report—anyone else, including the man who became her husband. (At last, in 1983, as if to atone for her earlier reticence, Innis published the first full review of scientific literature on homosexual behavior among animals, discovering references that included more than a hundred mammal species and numerous birds and reptiles.)[17]

Homosexual behavior is common enough among male giraffes, but that fact has been slow to emerge partly because of a reluctance on the part of scientists and science writers to report it or to describe the behavior as a sexual one. In a book published a decade after Innis's early report, one commentator describes male-with-male sexual behavior as "aberrant" and suggests that it could happen because "their reflexes become muddled."[18] The "muddled" concept seems outdated, but it is still fair to ask, Why do the males do it?

This is at once a behavioral and an evolutionary question. The behavioral answer I prefer is the simplest one. They do it because they choose to, and they choose to in response to some underlying desire or emotion or inclination. Perhaps we can say merely that the act gives them pleasure, in those circumstances, or possibly it provides a sense of satisfaction or completeness. The evolutionary answer requires us to presume that an act costing something in terms of reproductive success (added stress, a moment of vulnerability to predators, a loss of sperm) would probably be eliminated by evolutionary winnowing unless it in some contrary fashion improved reproductive success—or fitness. It could be, of course, that the fitness costs are not as great as they might at first seem. The moment of vulnerability to predators is brief. The loss of sperm, if any, would be insignificant, since males ordinarily produce an enormous surplus of it. Still, we ought to wonder about the possible fitness benefits of homosexual mating among giraffes.

Some of the benefits might include more reliable male bonds and alliances, or a more thoroughly defined dominance hierarchy. Developing a dominance hierarchy is fundamental, a very important part of the males' social lives. A gentle sort of necking can amount to friendly competition, but even in its friendliest or gentlest aspect, it may also serve to demonstrate power and therefore the potential to dominate. Necking among male giraffes, in other words, could help individuals figure out who can beat whom in a real fight, and thus could reduce the necessity of serious fighting and the potential damage that a serious fight can bring.

Serious fights do occur. As previously noted, a male's massive skull, provided with solid projections (the horns and secondary thickenings), makes a mean hammer that can be swung with great strength and amplified tremendously by that great neck working as a long, strong lever. The full weapon, functioning more like a medieval mace than a simple club, can produce devastating results. Two males in competition approach each other in a tall assertive posture, head on, legs stiff and neck high; if neither backs away, they close in, side by side, stern to bow, with each then boldly pressing the other until they start to trade long and swinging blows that, from a distance, may seem slow and graceful but can conclude with a deeply reverberating thud and an impact that communicates itself into the recipient's body as a tremendous shuddering of flesh and bone. The loser is driven away or knocked down, or even, on rare occasions, killed by the well-aimed blow.

—

Anne Innis left Fleur de Lys after completing her pioneering field study on the southern African giraffes. She traveled north, into Tanganyika and Kenya, looking for further opportunities to study wild giraffes in other populations and other regions, and to study other subspecies, but without success. She returned to Fleur de Lys after a few months, completing the giraffe film she had earlier begun, before, at last, leaving Africa and the wild giraffes. Forever, as it turned out.

In England, she married her fiancé, Ian Dagg, and together they returned to Canada, he to become a professor of physics at the University of Waterloo. They had three children, while she completed a biology PhD in 1967, producing a dissertation on the walking gaits of large mammals, including giraffes. She later studied camels and coauthored two important academic books on zoological subjects, *The Giraffe* (1976) and *The Camel* (1981). The former remains the sole significant monograph on giraffe biology, behavior, and ecology, while Dr. Anne Innis Dagg herself remains the well-recognized dean of giraffe studies. She never attained a full-time position in Canada as an academic zoologist, however. "I'd never give tenure to a married woman," one university dean explained to her in 1972. She relied upon nontenured, low-paying, part-time university teaching appointments for many years. Those unprestigious appointments meant in turn that she was forced to finance her continuing research on giraffes—as well as other species and nonzoological subjects—for the remainder of her career.

Innis's time at Fleur de Lys, altogether around half a year, was certainly not enough to answer all the questions about all aspects of giraffe behavior, or even to imagine what all the questions might be. Mothering behavior, for example, continues to be something of a mystery among giraffes, as does the more general issue of how giraffes relate to each other: that is, the nature of giraffe social groups or herds.

Innis began by concentrating on a handful of obvious behavioral categories, but even in her first weeks, the limitation of those categories became clear enough. She made her observations from inside her car, Camelo—except during the one hot day when, as she watched a group of giraffes about half a mile distant, she stepped out to stretch and practice ballet movements next to the car, steadying herself with the car's door handle. "Surely," she thought, "there's no harm in this." But then a female giraffe, drawn to Innis's strange behavior, slowly began to move, staring and coming closer, until she was within 40 yards (37 meters) of the car. Because she had not thought to establish a category for "curiosity" or curious "staring," the scientist was compelled to quit her stretching and get back into the car.[19]

Staring, of course, is a kind of behavior, and giraffe researchers would eventually incorporate it into the larger category of vigilance or vigilant behavior. Yet the curiosity that may underlie such staring is a mental state we can imagine far more easily than observe. Still, one can easily appreciate the likelihood that a variety of cognitive and emotional states—ranging from interest and a sense of ordinary wellness to curiosity to concern to genuine fear—may accompany the simple behavior of staring or vigilance. These cognitive and emotional states, though we may never directly observe them, must be essential for understanding how behaviors work in sequence. The behavior of flight follows the behavior of vigilance, for example, given the emotion of fear.

Concentrating on behavior in animals has long given scientists the luxury of avoiding a discussion about what causes behavior, and so the common public presumption has remained the one famously imagined by French philosopher René Descartes in the seventeenth century: that animals are biological machines operating through a series of mechanical instincts and reflexes. But it is a mistake to think of a giraffe as body disconnected from mind, to imagine *Giraffa camelopardalis* as a sophisticated yet mindless assembly of mere instinct and sheer reflex. Giraffes have minds or mental worlds, as we do—even though, to be sure, their mental worlds must be unique to giraffes and very different from yours and mine. What is the mental world of a giraffe? What is it like to be a giraffe? Those questions have yet to be asked and may be, in any case, unanswerable.

The opening photograph for this chapter documents a behavior Karl and I saw among Masai giraffes in the Masai Mara: what we call "leg-lifting." This seems to be an unusual form of sparring or fighting among males, a kind of giraffe jujitsu in which one male attempts to unbalance and topple the other with a surprise lift and twist of the leg. It appears that the behavior has not previously been photographed or written about. Karl subsequently photographed similar leg-lifting behavior among reticulated giraffes in the Samburu National Reserve.

The following four image sets illustrate three kinds of behavior that Anne Innis Dagg has described—eating, urine-testing, and "necking," or sparring and fighting (see Innis 1958)—along with the leg-lifting behavior.

Eating. Innis noted that giraffes use their tongues and lips to capture leaves before swiping or combing them off the twig or branch. The long tongue can be used to grasp leaves; the thick and hairy lips protect against thorns.

Urine-testing. Giraffe sex is preceded by a male or males testing a female to determine her state of fertility. The testing involves sampling her urine via the flehmen response: a flaring of the lips that opens a pair of dedicated scent ducts (the vomeronasal organ) in the roof of the mouth. In this instance, three males are following a female who may be fertile. It is likely that the males have already established a hierarchy defining who has priority.

Sparring and fighting. Two reticulated males square off, head to head, neck to neck, and then engage in a pushing contest that devolves into a series of hammer blows using the neck and head. The behavior can look graceful and even stylized from a distance; indeed, it is often hard to distinguish a gentle and at times eroticized sort of friendly sparring from a serious fight. A serious fight, however, will involve trading solid and potentially devastating blows. In the last image of this set, we see a Masai male landing a solid blow to knock his rival off-balance.

Leg-lifting. The final four photographs show in detail the sparring or fighting style we call leg-lifting, as demonstrated by two reticulated males at Samburu.

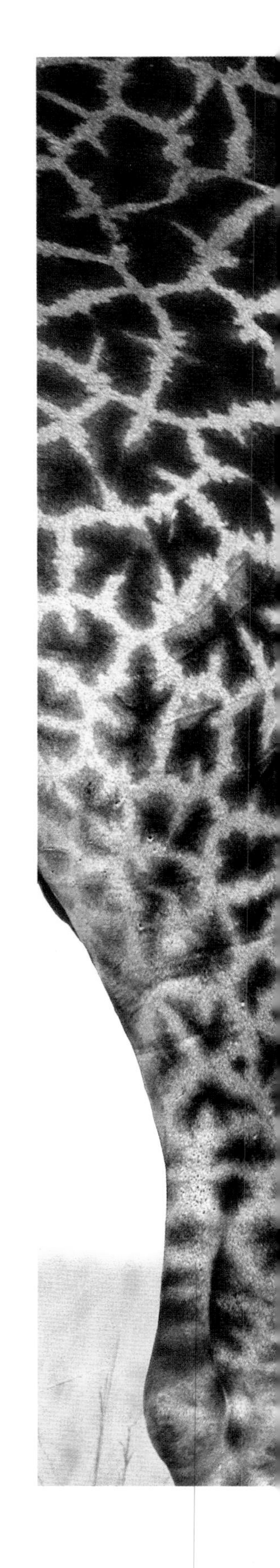

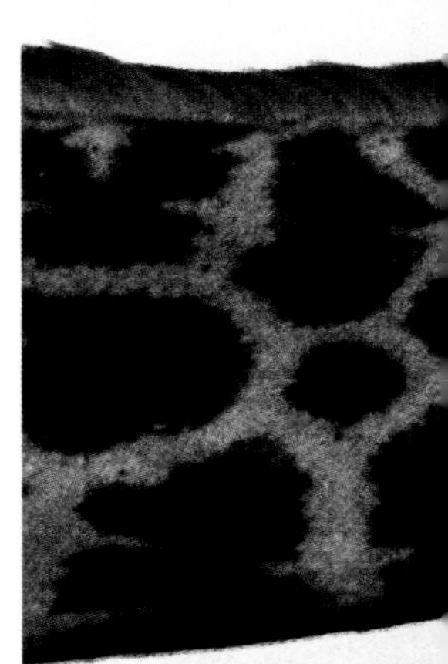

MOTHERS

AT THE FLEUR DE LYS FARM in South Africa, a fire set to clear brush and open new grassland for cattle unfortunately trapped three giraffes, including a mother and her infant. Leaping into thorn trees and flaring up in bright, ten-foot-high bursts of sparks, the flames were threatening the animals from three directions by the time zoologist Anne Innis and farm manager Alexander Matthew arrived.

The giraffes stood, apparently frozen in fear, as a worker, waving his arms, cursing and shouting, encouraged them to move: "Get out! Get out!" he shouted. "That way, you stupid brutes!"

Only when the fire was near enough to burn them did the three animals shift into action, suddenly racing through the flames and over to an already burned spot of grass. Away from the fire now, the giraffes appeared to calm down: the infant (six feet tall but wobbly and fragile, with the umbilical cord still attached) pausing to nurse, the mother lowering her head to reassure her baby with a gentle touch on the back. Soon, though, the nursing and reassurance were over. The mother ambled over to a fence at the edge of the burnt area, then stepped over it. The baby followed but was unable to continue past the barrier. The mother strode away for a moment, then returned to look at her baby on the other side of the fence, the pair of them now simply staring at each other. At last, the mother turned and strode off into a thicket, leaving her infant behind.

For Innis, the maternal drama was anguishing. "Doesn't she realize the baby can't get over the fence?" she asked Matthew.

"Giraffe aren't very smart," he responded.

"What'll we do? Can we cut the fence?"

"No. That would scare both of them away and, anyway, it's getting dark. We'll have to leave them. Probably the mother will come back later."

Innis spent the evening wracked with concern, believing that a newborn giraffe could not possibly survive alone for more than a few hours. Would the mother really return? Why had she left to begin with? The young researcher drove back to the burned-over area at dawn, hoping to see some signs of either one, but she found nothing. She never learned what happened to the mother or her baby.[1]

—

Do giraffes make bad mothers?

That may seem an odd question. Since good mothering is directly related to reproductive success for females, evolution should inevitably prime any species to produce the best possible mothers given that species' ecological circumstances. In the case of mammals, mothering behavior based on an emotional attachment or bond between the mother and her young offspring is particularly important because most newborns and youngsters pass through a highly vulnerable period when they depend directly on their mothers for both nutrition and protection from predators.

Moreover, giraffes are not prolific reproducers, so that each new baby ought to represent a major genetic investment for the mother. Females are likely to become first-time mothers by the time they reach five years. Given a gestation period of around fifteen months, they are maximally capable of producing a new baby every eighteen months, on average: a theoretical maximum of around eleven offspring altogether by the time their reproductive careers end around the age of twenty. Eleven offspring under ideal conditions is not a large number, and they are very vulnerable to attack by some predators—mainly lions, hyenas, and leopards. One would imagine, then, that a giraffe mother would not only be ready to nurse during her youngster's early months; she would also be intensely attuned to protecting her young offspring from dangerous predators. How unsettling, then, that the earliest scientific accounts of giraffe behavior argued the reverse to be true.

Anne Innis Dagg, who in 1956 became the first person to conduct a field study of wild giraffes and who later spent eight months studying a captive herd of eighteen giraffes at the Taronga Zoo in Sydney, Australia, expressed that conclusion in her first scientific publication, in 1958.[2] J. Bristol Foster, a Canadian zoologist who taught at the University of Nairobi, acquired the same perspective after studying wild giraffes from 1965 to 1968 in Kenya's Nairobi National Park.[3] The two experts jointly declared in their coauthored book *The Giraffe: Its Biology, Behavior, and Ecology* ([1976] 1982)—to this day, the sole authoritative monograph on the species—that the bond between giraffe mothers and their young is "remarkably weak."[4]

Part of the evidence for this surprising assertion comes from the zoo literature. There is, for example, a summary written in 1925 by Sol Stephan, general manager of the Cincinnati Zoo, titled "Forty Years' Experience with Giraffes in Captivity." The zoo's experience included, in October of 1889, the first birth of a giraffe anywhere in the Americas. The baby lived only six days, Stephan writes, his death "primarily due to the fact that the mother refused to nurse it and would strike at the young one whenever it came near her."[5] Another sad tale, written in 1966 by James Savoy, superintendent of the Columbus Zoo, describes a female giraffe who "trampled . . . to death" her firstborn and thereafter "repeatedly abandoned her calves."[6] Then there is a 1955 report from the San Diego Zoo of a newborn born unable to stand because of an injury that occurred during birth. In this instance, to quote Dagg and Foster, the infant was promptly "abandoned" by his mother.[7]

These and other zoo accounts seemed to support a vision of giraffe mothers as particularly shy, recalcitrant, uninterested, and all too likely to abandon their offspring for reasons that may or may not be comprehensible to the human observer.[8] In fact, however, the 1955 report from the San Diego Zoo, written by zookeeper George H. Pournelle, actually describes the reverse of abandonment. According to Pournelle, the mother nuzzled and licked and was "quite solicitous" of her injured newborn, and she seemed to resent anyone approaching her enclosure. Concerned that the injured infant was unable to stand into a nursing position, though, the zoo management finally separated him from the mother—their choice, not hers.

The infant, placed in his own sleeping compartment at the zoo, died of pneumonia early the next morning.[9]

The 1966 report from the Columbus Zoo does claim to describe the case of a mother who "repeatedly abandoned her calves" after killing the first, but it then goes on to reveal that zoo management chose to remove her next three offspring immediately after they were born, without waiting to see whether she would abandon them or not. Two of those three then died due to inadequate husbandry.

And Sol Stephan's 1925 summary of four decades' experience in giraffe keeping at the Cincinnati Zoo does indeed assert that the mother of America's first captive-born giraffe failed to nurse her infant and also threatened him with kicks. The Cincinnati account also documents that the zoo gave its giraffes an entirely unnatural diet (beets, cabbage, carrots, onions; stale bread and crackers; crushed oats, red clover hay, and wheat bran), which they were required to eat in an unnatural position (bending down to ground level), while living in unnatural confinement (a cage) under unnatural social conditions (the zoo having started its giraffe "collection" with the random acquisition of one male and one female in 1878).

Yes, the zoo literature cited by Dagg and Foster includes a handful of accounts of giraffe mothers who accidentally stepped on their infants or refused to allow them to suckle or even seemed to show direct hostility—deliberately, it would seem, kicking or even trampling them.[10] We cannot be certain, however, whether these examples are anomalous or representative. And it will never be clear whether they are evidence of an abnormally weak bond among giraffe mothers and infants or—what is equally likely—evidence of the abnormally stressful and impoverished environments historically offered by most zoos.

—

By the time Dagg and Foster had finished their book, scientific field studies of giraffes were still just beginning. As a result, the principal piece of evidence from the field supporting the weak-bond hypothesis came from Foster's own regular observations at Nairobi National Park of an adult female standing "together with more than one newborn."[11] He would sight as many as five very young giraffes together in a group, and sometimes they would be accompanied by only a single adult female—or even none at all. Calves as young as a month old could often be found more than a mile away from their mothers, and it seemed to Foster as if nursing ended altogether after about a month.

To further complicate the picture, an anecdotal account published anonymously in a 1965 issue of *African Wild Life* magazine relayed the experience of two visitors to Kruger National Park who watched nine adult female giraffes arrayed in a circle. After ten minutes, the gathering spread out enough that the observers could see that a baby had just been born. The infant struggled to get to his feet for more than half an hour before finally standing up fully, then gamboling tentatively around his mother, after which all the "midwives" (in the language of the account) moved in to nuzzle the baby. Three male giraffes had been watching these events from a distance, never moving closer than about fifty yards, while two hyenas observed, hungrily perhaps, from a greater distance.[12]

In 1969, meanwhile, a Colombian zoology student associated with the Max Planck Institute in Germany and the Serengeti Research Institute of Tanzania began an extended study of giraffes in the Serengeti National Park. Driving a Land Rover around his 100-square-mile (161-square-kilometer) study area, Carlos Mejia came to recognize some 350 individual animals. His technique was similar to that pioneered by J. Bristol Foster—developing a photographic catalogue based on spot patterns at the neck—but his work was a good deal more ambitious. Two or three days of the week, Mejia drove through an established route of the study area and recorded which individuals he saw where, also noting who was with whom whenever he came upon a group. The remainder of his time was spent following individuals or small groups for a half day, a full day, or even (when the moon was full) a day and a half. These follows enabled Mejia to assess behavioral patterns; he gathered data in the form of notes, taken at three-minute intervals, of what each animal was doing.

While Foster had conducted a general study of giraffes, Mejia focused on relationships between mothers and their young. He found that mothers who had just given birth remained briefly isolated from the herd, deliberately secluded, so it seemed, with their newborns for the first few days. Mejia described one mother who attacked any other female attempting to approach during that critical seclusion period, which was a time, the researcher noticed, when a mother frequently nuzzles and licks her calf. Perhaps this period of self-enforced isolation was necessary for the mother and offspring to develop mutual recognition and attachment.

After those initial days of seclusion, Mejia reported, the mother would lead her calf into an area where there were other mothers and other tottering, six-foot-tall babies. Mejia believed that the mothers sought each other mainly because of their calves, and that the calves joined together—or bonded—more strongly than the mothers did with each other. (Trying to measure emotional bonds by considering physical distances, Mejia found shorter average distances among the calves than between calves and their mothers.)[13] He was struck, also, by how often the youngsters would approach each other nose first, touching noses and sometimes licking each other's noses before suddenly leaping apart. It was a strange kind of playful nursery—or, as an earlier commentator described it, a "kindergarten."[14]

The idea of a giraffe kindergarten was originally suggested by a German researcher named Dieter Backhaus, who briefly studied giraffes in the Garamba National Park in northeastern Zaire (now the Democratic Republic of the Congo).[15] Backhaus had considered such a social arrangement to consist of a group of youngsters temporarily watched over by a single adult female. Mejia, however, found that the young animals were often left completely on their own, typically out on an open hilltop. The mothers would wander off, browsing in some hillside thicket or woodsy gully, moving a half mile or even a mile away before returning near the end of the day, locating their own young, and nursing them. It must have seemed a little like mom going off to work, since the mothers left their youngsters at around nine o'clock in the morning, as the shadows shortened and the day warmed, then returned at around five o'clock in the afternoon, as the shadows began to lengthen again and the evening breezes moved in. Giraffe calves in the kindergarten would spend much of the day lying around, occasionally nibbling at various items of food—demonstrating a gradually growing tolerance for solid foods that began not many days after birth.[16] Still, while Dagg and Foster had earlier concluded that giraffe infants were weaning within a month, Mejia reported that the animals he followed in the Serengeti continued to nurse for about a year. In the kindergarten, they usually nursed once in the evening, when their mothers returned, and then again in the morning before the mothers departed.

Why a giraffe kindergarten? It is possible that this system evolved because it enables the mothers to seek the richer sources of nutrition, to go where the leaves are, in the thickets and woodsy gullies, while leaving their youngsters hidden in the grass on a hilltop, where there is less likelihood of a predator sneaking up on them. Predation is indeed a serious matter, with lions and hyenas probably accounting for a significant part of the 50 percent infant mortality rate (for giraffes up to 3 months old) that Mejia found in the Serengeti, or the 73 percent infant mortality (for giraffes up to one year old) that Foster reported from Nairobi National Park. But with the mothers away, does the kindergarten actually provide much protection for the young giraffes?

Mejia's single observation of a predation attempt showed how important mothers could be in defending their young. He watched a lion jump a calf with the mother standing by; the mother immediately began to kick, which temporarily drove the lion back, whereupon mother and young ran away, the lion following but finally unable to keep up.[17]

—

Carlos Mejia's results, based on giraffes in the Serengeti, were reinforced by an important study conducted by Vaughan Langman, a zoology graduate student from Alaska. Langman arrived at the Timbavati Nature Preserve in South Africa (as it happened, only about ten miles [sixteen kilometers] from where Anne Innis had watched giraffes in 1956 at Fleur de Lys), intending to do his doctoral work on one of the several small mammals of the region. Just about any species would do, he told himself, and smaller animals seemed simpler and easier to deal with than bigger ones. But then he became interested in giraffes. "When you observe giraffes closely for a long time," as he would later write, "you cannot help being impressed by the height, body size and grace of these animals."[18]

They were attractive and fascinating, but Langman also began to recognize that the scientific literature, what there was of it, indicated that giraffes had a poorly understood social system and an unusual physiology. Giraffes would certainly be interesting for someone oriented, as Langman was, to issues in physiology, and he found himself wondering about the cooling problem. How did giraffes keep themselves sufficiently cool under the hot African sun? For a large animal to keep cool by sweating, as humans do, would require too much water, he thought. And for a giraffe to cool by panting, as a dog does, would probably not work either, since it would involve moving a good deal of hot air up and down that long windpipe, a process that would create more heat than it dissipated.

Then there was the "poor mothers" problem. Langman read the scientific literature describing giraffe females as being skittish, unreliable mothers who weaned their infants after only a month, leaving them to fend for themselves when they were still highly vulnerable. How was that possible? How were these fragile calves able to find sufficient food and avoid the predatory tooth and claw? And why would evolution ever produce such a seemingly weak or even self-defeating system in the first place?

Langman thought he would be able to answer such questions after about a year of initial research. He would begin, he imagined, by learning to recognize individual animals by their spot patterns. Then he would make sense of how they reacted to one another: their social system. That may have seemed like a good plan, but it proved to be vastly optimistic. The few individuals Langman first came to recognize would wander into a thicket and disappear for several weeks.

The young zoology student decided to try placing radio transmitters on the animals. To keep the transmitters from being broken during the ordinary scrapings and bangings of a giraffe's life, he would embed them in fiberglass. He would attach the transmitters to leather collars that had been softened, in order to avoid chafing the animal's neck, with a layer of glued-on horse blanket. And finally, to make sure the collar did not remain on much longer than the study lasted, he would complete the leather loop with brass bolts that, he hoped, would break off or wear away within a couple of years. Meanwhile, it took six months of preparation before Langman was confident enough to try placing collars on wild giraffes. Collaring was a complicated and dangerous procedure that required darting with just the right amount of tranquilizer, sufficient to take down a 1,500 to 3,000 pound animal, for just enough time to attach the collar.

Eventually, and with the help of a dozen assistants, Langman captured and collared enough giraffes to constitute a useful study group, and so he began his observations: going out in a Land Rover at dawn, returning to his camp at dusk. At first, his observations were confusing and, all too often, boring; he was still generally uncertain about who was doing what where and with whom. But the radio tracking helped him learn to recognize individuals visually from a distance, based on the shape of a head, the style of horns, the manner of a gait. And in that way, finally, he came to know his animals well enough that watching them turned into a highly entertaining experience. It was like watching a daily film series. "Now leaving camp in the morning was exciting, and I couldn't wait to see how the story or series of events would turn out. Where would our giraffes go today? What other giraffes would be encountered on the way?"[19]

Langman's research showed that giraffe mothers did not wean or abandon their young after a month. It also directly challenged the ideas that giraffes were bad mothers and that the bond between mother and infant was, somehow, "remarkably weak." To appreciate the nature of giraffe mothering more fully, Langman came to believe, requires first recognizing that hoofed animals in general have evolved to use two different approaches for protecting their young: following and hiding. Species adapted to the following strategy produce youngsters who quickly learn to follow their mothers: wildebeests, for example. Newborn wildebeests struggle to their feet and within a few hours start to follow their mothers, while their mothers soon return to their usual habit of following the herd. Follower species find safety in numbers.

Hider species—Grant's gazelles, for instance—keep their infants hidden immediately after birth, with the mothers moving away from their offspring, ideally deceiving any potential predators, and only returning to nurse. During this early lying-out period, the infant moves little and spends most of his or her time lying down: a shadow in the shivering grass. The mother may return two or three times each day to nurse, but otherwise she wanders away to feed. A giraffe mother will wander as far as 15 miles (about 24 kilometers) distant during a day, and when she does return, her young offspring is primed not to leap up and run to her immediately, but instead to remain lying down, allowing the mother to approach, then to nudge and lick, before beginning to nurse.

This early lying-out phase, the strategy of a hider species, ordinarily ends after one to four weeks, whereupon the mothers and their offspring join the herd. For giraffes, though, the young then join what Langman called "nursery herds," comparable to the giraffe kindergartens Dieter Backhaus referred to. Nursery herds form in the early morning, according to Langman, as the mothers lead their infants to particular areas, often comparatively open and on a hilltop. The mothers and their young will all feed in this area for a brief while before the mothers depart, one by one, until only the young remain. They lie down or occasionally nibble at leaves, and twice or thrice during the day the mother returns to nurse before leaving once more. As the daylight fades, though, the mothers amble back, and as darkness falls, most of the animals in this group lie down. They fold their legs beneath their bodies. They curl their necks in order to rest their heads on their haunches. They sleep. But one or perhaps two of the adult females from this group will remain standing, sentries on the watch for predators; and when they become tired and lie down, others in the group rise to their feet and take over the watch.

Sometimes all the adult giraffes in this nursery group leave during the day. Other times, Langman found, a single mother might remain, keeping watch as a sort of babysitter and protector. What makes this mother stay behind during the day? Langman believed the likely answer was that she felt no strong urge to drink or wander about in search of food. Having no compelling reason to leave, she stayed—watching over her own calf but incidentally protecting the others as well. Should she become anxious at the sight of an approaching predator, she finds her own youngster and urges him or her to follow. She runs, the calf runs, and soon the other calves, recognizing the anxiety of the mother and witnessing the flight of the pair of them, also run. Thus, the mother, in attempting to protect her own young, may happen to protect the others as well.

So the original idea, that giraffe mothers abandoned their young after a short while, was based on some incomplete observations. Better observations showed that the mothers were actually hiding their young. But why were they doing it for such a long time? These giraffe nursery groups could last for up to a year, while all the other species known as hiders only kept their youngsters under wraps for a month at most. It was extraordinary that giraffes extended the period of hiding for so long, and why they did so, Langman thought, looked like a deep puzzle without any clues.

During a conversation over a campfire one evening, a visiting friend asked Langman how giraffes were able to drink, given that their long necks seemed not long enough to reach water at their feet. Langham said that drinking from a pool or stream required an awkward sort of bending or spraddling of the legs. The awkwardness of that position, of course, made drinking at a waterhole dangerous for any giraffe, but especially so for the already vulnerable young. The guest then asked whether the youngsters also drank in that fashion, and Langman realized that he had never actually seen a young giraffe drink. Were they only drinking at night? After some time spent observing watering holes at night, Langman concluded that they were not drinking at night, either.

Giraffes derive water from the plants they eat,[20] and of course the youngsters nurse as well, but, given the intensity of the overhead sun and the heat of the day, how could the young ones survive without drinking any water at all? Langman suspected that the answer to that question had to do with their radically extended period of lying-out, the long-term nursery groups, but in order to consider such questions and intuitions further, he needed to know more about how much water a giraffe needs in the first place, and since an animal's water requirements are related to temperature regulation, he felt he should also learn more about how giraffes maintain their normal body temperature.

Langman—by now Dr. Langman—moved to Kenya and started a new research project, this one in collaboration with Kenyan physiologist Geoffrey M. Ole Maloiy, that examined the heat-regulation physiology of giraffes. The project involved capturing some wild giraffes in order to monitor, in various ways, their metabolism and body temperatures. Most animals stabilize their body temperatures through active means, such as panting and sweating; but, as the research by Langman and Maloiy demonstrated, giraffes simply allow their internal temperature to fluctuate, and it will shift by anywhere from 5.4 to

18 degrees Fahrenheit (3 to 10 degrees Celsius) on a typical day. Camels do the same. In this way, giraffes, like camels, are able to reduce their water requirements by a great deal. A large body makes this trick possible. Giraffes and camels are big enough that their body mass provides a high level of thermal inertia, buffering the extremes in air temperature, gradualizing the changes in body temperature.

This made sense for giraffe adults, but what about the babies and the young ones who were not especially large? One day, as Langman was keeping himself cool by sitting in the shade of a thorn tree, he noticed that a young giraffe in the experiment, ordinarily reluctant to approach, had moved up close on the other side of a fence in order to find a bit of shade for himself. "Then it occurred to me: the last time I had seen a young giraffe behave like that was when I was watching them in a nursery herd."[21]

Why does the hiding period last so long? According to Langman's thinking, giraffes were not merely staying in nursery groups to hide from predators. They were also, during this extended period, hiding from the sun—in the tall grass, in the shade of trees and bushes—until they were big enough to acquire the necessary body mass and, thus, thermal inertia.

—

During her recent study of Rothschild's giraffes at the Soysambu Conservancy in Kenya, British researcher Zoe Muller witnessed an event that made her reexamine everything she thought she knew about giraffe motherhood.[22]

The giraffes in the Soysambu Conservancy are fortunate to be living in an area where lions have been exterminated. Bad for lions is good for giraffes, and Muller reports that the giraffes there are abundantly reproducing and have a comparatively low mortality rate. One giraffe, known in the research project as F008, was especially notable because she had given birth to an infant with a badly deformed rear leg. The pair was thus easy to identify. Would she "abandon" this crippled offspring the way zoo giraffe mothers supposedly did? On the contrary, F008 seemed especially attentive, even, in Muller's assessment, "doting" as a mother. At four weeks, the calf was able to stand and nurse, but he had great difficulty walking. Thus, he spent most of his time standing in one place, while the mother stayed very close to her compromised infant, never moving (according to each of Muller's many observations) more than about 22 yards (some 20 meters) away and remaining invariably vigilant, watching for trouble or danger no matter what the rest of the herd happened to be doing. This level of maternal closeness and attentiveness was unusual, since, as we have seen, the ordinary system was for mothers to hide their calves during the first four weeks of life and wander away in search of food. So F008 was sacrificing her own well-being—probably going hungry, certainly reducing her foraging opportunities—in order to stay close and watch over her handicapped baby.

On May 4, 2010, as Muller was driving her standard route along the research transect, she came upon a herd of seventeen giraffes, all of them female. It was an unusually large herd in an area where she rarely found giraffes, but even more unusual was their behavior. They were all "highly vigilant," according to Muller, and "running around in apparently bizarre patterns."[23] They also seemed afraid of her vehicle, which itself was very strange. Normally, the Soysambu giraffes were unfazed by the approach of her car; they were entirely habituated to it.

The researcher stopped to watch, anticipating that the seventeen females would become less agitated over time, yet they continued with their seemingly distressed behaviors. Muller eventually recognized that they seemed to be focusing their attention on a particular spot in a thicket. They would run over to that spot, stare, and then run away. Muller drove over to the spot and found, in an open, grassy area, the body of F008's crippled calf. After examining the carcass, she concluded that the four-week-old animal had died quite recently from some kind of natural cause. The mother, indeed, was one of the seventeen females running around.

Muller drove away from the carcass and parked in a place where she had a good view of the area. She stayed for the next three hours and watched, reporting in her notes that "all 17 female giraffe ran around the area being vigilant, continually approached / retreated from the carcass and showed extreme interest in it."[24]

When Muller returned that afternoon, she found the herd had expanded to include twenty-three females and four juveniles. The animals were still "quite restless" and "being vigilant" in the direction of the carcass. Moreover, the adult females in this group were approaching the carcass and "'nudging' it with their muzzles, then lifting their heads to look around before bending down to nudge it again." The juveniles, meanwhile, were also examining the carcass, coming close enough to bend down and sniff before leaping up and running off. And when Muller returned that evening, she discovered fifteen adult females, including F008, all of them now "clustered" around the body.

The researcher imagined that scavengers—hyenas, perhaps—would come during the night and remove the body, but when she returned on the morning of the second day, it was still there, in one piece and surrounded now by seven adult females, one of whom was the mother. All seven walked around the body and were highly attentive toward it. That afternoon, Muller counted a total of fifteen giraffes in the area, four of them adult males. The females continued to show a great interest in the carcass, regularly approaching and inspecting it, while the four males showed absolutely no interest. They were either eating or approaching and inspecting the females. Muller returned again at 8:45 that evening, finding this time three females, including the mother, all three of them standing no farther than 10 yards (about 10 meters) from the carcass, highly attentive to their surroundings and the body, and neither foraging nor ruminating.

When Muller returned on the morning of the third day, she found only the mother standing vigil around the body, while the body itself had been half eaten by a nighttime scavenger and dragged for a considerable distance from its original position. The mother stood directly over the remains that morning, and she was still there when Muller came back in the afternoon and again in the evening. "She was just standing there being vigilant, and not foraging or ruminating."[25]

On the morning of the fourth day, the mother was still in the area, although she now stood about 220 yards (200 meters) away from the spot where the half-eaten carcass had been the day before. When Muller drove over to that spot, she found that the young giraffe's remains were entirely gone, apparently finished off during the night by the skulking scavenger. When Muller came back again that afternoon, the mother was gone too.[26]

The lead photograph for this chapter presents a simple portrait of motherly love. The following photographs begin with a late-day look at the start of a birth. The mother is standing; the newborn's front feet appear first. The next two images show, in the early morning light, a mother and her newborn (who is perhaps just a few hours and at most one or two days old), along with a second adult who seems to be a curious female onlooker. Note the string of saliva extending from mother to young; she has been licking her newborn.
The next two photos may suggest (falsely) a very young giraffe who is lost or has been abandoned. An image of what could be a giraffe "kindergarten" or "nursery group" is next, followed by a look at two youngsters in a nursery group, both trying to nibble a bit of grass. The images end with a five-photograph meditation on the difficulties and pleasures of nursing, and a final nod to the glowing beauty of maternal attachment, or motherly love.

OTHERS

GIRAFFES ARE GREGARIOUS BUT NOT TERRITORIAL. "Not territorial" means that no one tries to own real estate. "Gregarious" means that they often spend time in each other's company.[1]

Giraffe social groups, or herds, commonly consist of roughly a half dozen individuals, with more precise averages showing considerable regional variance.[2] Herds can be as small as two individuals moving together or as large as twenty or thirty or even fifty and more. One researcher counted 175 giraffes gathered in a single place.[3] But these groups are also remarkably unstable. Any herd will vary in size and composition from day to day, hour to hour, or even moment to moment. Giraffes come and go, moving into and out of the herd, associating with others and moving in the same direction at one time, shifting directions and splitting associations at another. Some scientists write about the instability or fluidity of giraffe society. Others have described the social system as one of fission-fusion, evoking the image of groups splitting apart, then fusing back together in an irregular pattern.

Among the primates, chimpanzees are often described as a fission-fusion species, in which (within a larger community of individuals who know each other and, unlike giraffes, inhabit and defend a clearly defined territory) groups of individuals form and disband as the day goes on. But perhaps we can recognize the fission-fusion style most clearly in our own lives. We humans routinely spend our daily lives moving from group to group, with the groups varying in size and forming and dissolving in a pattern complex enough that it might seem random to, say, the casual alien observer from outer space. Ours is not at all a random social system, of course, since we move with purpose and predictably seek out the company of family, friends, and familiars.

The fission-fusion society of giraffes is not random either. Some of that nonrandomness I described in the last chapter. Mothers form bonds with their nursing offspring; members of a nursing group tend to stick together as playmates. Swiss researcher Barbara Leuthold found, in a 1970s study of giraffe society in Kenya's Tsavo East National Park, that the calves in a nursery group often playfully ran and jumped about, although the males, starting at around the age of one month, began associating with one another by play-fighting, or sparring with their heads and necks—what I have earlier referred to as "necking."

Yet the strong associations between mother and young seemed to end a few months after nursing did. Leuthold believed that mothers defined when nursing took place and for how long. A mother would approach her young. When she came close enough—to within a few yards—the juvenile would look up, recognize the mother, and move in to suckle. There were no obviously audible calls, so perhaps the mother and her offspring recognized each other by sight and, secondarily, by smell. Leuthold frequently identified nursing juveniles who were nine months old and, on rare occasions, some up to thirteen months. But the fifteen juveniles in her study group who survived the first year of life continued to associate with their mothers for two to five months after weaning. So giraffes could be dependent on their mothers for as long as a year and a half; then, after leaving their mothers, they moved into what Leuthold called the subadult (or adolescent) stage.

Among adult females, Leuthold found some consistency in individual associations: predictable pairs occurring when at least one of the adults was also accompanied by a juvenile. Other than the pairing of mothers, though, it seemed that adult females rarely interacted with one another.

Among the males, she discovered a more pronounced variation in sociability, with the adolescent males being highly gregarious. They regularly associated with each other in groups and, conversely, were sighted alone less than 5 percent of the time. In each other's company, the adolescent males frequently engaged in a relaxed sort of sparring or necking that, on occasion, culminated in mounting with an erect penis. This generally social phase of a male's life, she concluded, was a time of testing his relative power within a gang of peers, of establishing his place in the male dominance hierarchy.

Once they left adolescence, the young adult males were no longer quite so gregarious. Young adult males were seen alone from 5 to 15 percent of the time, and they no longer playfully interacted through necking. Still, the adult males recognized each other as individuals, and, according to Leuthold, they clearly understood who was dominant over whom ("an absolute hierarchy existed among them").[4] Adult males and females would usually intermingle randomly, with little evidence of mutual interest, but when a female entered her fertile period, one of the adult males would guard her from the other males. Should any subordinate approach, the guarding male, head and chest held high, would walk directly toward the intruder, who then, typically, would lower his head and quietly move off or gracelessly scramble away.

For a male, getting older—passing out of the young adult stage and into what Leuthold called the fully adult stage and, from there, to the geriatric category—was marked by a steady decline in gregariousness. The fully adult males were discovered to be alone in 15 to 30 percent of her sightings, and the old males were sighted alone from 30 to 50 percent of the time. The fully adult males often associated with females, driven to do so for sexual reasons, while the old males never sexually approached females at all. And even when they moved into the company of others, male or female, the old fellows still acted remote.

—

Barbara Leuthold's research may leave the impression that giraffes form only passing and shallow relationships, with the most important one—between mother and nursing offspring—ending not long after the mother stops being a source of milk and protection. Other scientists following Leuthold, however, have developed a more detailed and complex picture of giraffe society.

We now recognize that the fission-fusion society of giraffes, with herds forming and disintegrating, takes place within the larger context of a stable geographical community maintained by the females. The younger, more vigorous adult males wander across home ranges that can be quite large. Their home ranges are nonexclusive: that is, the males do not compete with one another over territory. But it is the adult females, not the males, who provide geographic stability. They tend to remain in a particular area, and their stable presence essentially defines the larger community from which the fission-fusion herds derive. The unstable herds, then, are subunits of much larger, less visible, more stable social networks.[5]

These larger communities exist and are maintained over time because the giraffes themselves maintain long-term emotional bonds with one another. The idea that giraffes have long-term bonds of any sort is indeed a relatively new one, although a few researchers imagined that it might be so. Australian ecologist Julian Fennessy, who has studied giraffes in Namibia and elsewhere across Africa for more than a decade, reports seeing, in one population, the pattern of particular adult females showing up in the company of the same others around a third to half the time. In another population, this one mostly males, Fennessy noticed a similar pattern of social preferences among males. So, in his words, it has begun to look as if the Namibian desert-dwelling giraffes are living in societies of "close-knit 'friends,' so to speak, plus some other giraffes that come along for a while and then move on."[6]

Friends? Since it is clear that giraffes recognize each other as individuals, it is fair to ask if they seek out particular individuals to spend time with. If they prefer the company of some individuals over others, are these preferences significant? Do giraffes form only temporary and casual bonds, or do they also maintain stronger ones that come closer to what we humans think of as friendships?[7]

Psychologist Meredith Bashaw began thinking about the emotional bonds between giraffes after she learned that the Atlanta Zoo in Georgia was about to break up a trio of giraffes—two females and one male—who had spent nine years together in the same enclosure. The zoo's management intended to trade away the male, and Bashaw wanted to see whether the two females would show any distress at their long-time companion's disappearance. In fact, their emotional reaction was surprisingly "huge."[8] Both females responded with endless pacing and obsessive licking of the enclosure fence, both clear signs of psychological distress. After ten days, one female's unnatural pacing began to diminish, while the other continued to pace and lick the fence until the zookeepers finally intervened, successfully diverting the two animals by introducing new challenges in their food dispersal system.[9]

The prevailing view had been that adult giraffes are socially aloof and emotionally indifferent to one another. But the clear distress shown by those two female giraffes after the disappearance of their long-term social companion (the male had never mated with either) suggested to Bashaw a completely different picture, one that she went on to study more fully in California.

Perhaps the most natural setting for captive giraffes anywhere can be found at the San Diego Zoo's Wild Animal Park, which maintains a 90-acre (36.5-hectare) habitat for East African animals: rolling grassy savannas interrupted by a few clustered palm trees, a couple of large ponds, a stream, and a rock-face canyon, all inhabited by eleven ungulate and two bird species. When Bashaw began her two-year research project in 2002, the East African exhibit included a dozen Rothschild's giraffes, six of them adult females, the remainder a varying mixture of adult males and immatures of both sexes.

Her method was simple enough. In order to look for long-term social preferences, she made daily surveys, five days a week, in which she focused for twenty minutes on the behavior and social circumstance of each of the dozen giraffes. During that twenty minutes, she identified at one-minute intervals two different social criteria—neighborhood proximity and close proximity—and kept a full record of any affiliative behaviors. Neighborhood proximity answered the question, Who is the nearest giraffe? Close proximity identified any nearest giraffe who happened to be within two neck-lengths of the focal animal. Affiliative behaviors referred to friendly ones and included such actions as approaching, necking, head rubbing, bumping, social examination through sniffing or licking, muzzle-to-muzzle contact, and eating from the same feeder or part of a plant.[10]

After two years of gathering data, Bashaw subjected her results to statistical analysis, which clarified that the associations between individual giraffes at the San Diego exhibit were not random. Combining those analyzed results with known information about family relationships further clarified that these nonrandom associations were not friendships between age peers but rather relationships among relatives: more specifically, mother-daughter associations that had persisted after weaning and continued into adulthood. A prolonged bond between mothers and daughters, then, seemed to make up the important lines of social connection among these animals. Bashaw's study of zoo giraffes appears to confirm observations made in one or two field studies,[11] while the continuation of strong mother-daughter relationships into adulthood has also been described for other ungulates, including sheep, domestic cattle, and bison.[12]

Mothers rule. The matrilineal lines of force in giraffe society struggle in a dynamic embrace with shifting food resources and other ecological circumstances to regulate the movement of individuals across time and space.

—

A giraffe's meaningful universe is largely a social one in which the important other is another of the same species. This attentional bias, which I might call species narcissism, characterizes any species capable of attention: It marks the human experience as much as the experience of giraffes. But beyond that meaningful social universe, there are still others of other species, and they can be experienced as positive, negative, or neutral.

For giraffes, negative others include several kinds of tick parasites. The ticks' irritating and unhealthy activities, fortunately, are routinely counteracted by two kinds of positive others: the tickbirds, otherwise known as red-billed and yellow-billed oxpeckers (both species of the *Buphagus* genus), who spend a good deal of their day riding on the neck, head, or back of giraffes, or any other part that may be readily available, while hopping hither and thither and pecking away at ticks. The birds thus feed themselves and get a free ride in a safe place, while in return reducing the parasite load for their giraffe hosts. This is an instance of intraspecies mutualism, a relationship providing benefits to both picker and picked, although it is never clear that the giraffe has any sense of the bird except as a hopping, picking, fluttering constant.[13]

The most obvious other-species threats come from predators, including cheetahs, leopards, hyenas, and lions, with the last being the most feared and serious. Mortality from predation of all sorts is surprisingly high, from 50 to 70 percent,[14] especially among the vulnerable young. Giraffes sometimes congregate with other savanna browsers and grazers—zebras, wildebeests, elands, and so on—and such multispecies congregations may amount to seeking safety in numbers. Forty eyes are better than twenty, after all, and a single diner's appetite remains the same no matter how large the menu.

Although some experts have debated whether the giraffe's distinctive coloration and coat pattern serve as camouflage against predators, my own experience suggests that it does. That uneven, stick-and-thorn-and-leaf pattern in dry browns and dull creams becomes effective whenever this creature feeds in a cluster of acacia trees or moves into a complicated thicket. On the open plains, giraffes act as if they understand that their camouflage is useless and their unique profile hard to miss. Nothing to be done. They show no impulse to hide or freeze or otherwise blend into the surroundings, and so, it would seem, their defense shifts from disguise to vigilance. An approaching anomaly, whether threat or potential threat, brings them to look up and gaze intently. Good hearing, excellent vision, and a high perspective make this kind of passive defense entirely reasonable. And should the potential threat continue to approach, they can turn and actively evade, galloping away with enough speed to put them quickly beyond the teeth and claws of any lion.

A final defense is kicking. If a giraffe has been surprised or threatened by a leopard or cornered by lions or hyenas, or if a mother has stopped to defend her still-wobbly infant, then she will turn to face the source of danger. She can strike forward with her shins or her heavy, wedge-shaped front hooves, or she can turn about and kick back with her hind legs to render a solid blow to the rear. One observer has described a mother successfully defending her young by maintaining a steady barrage of kicks until the predator at last gave up. Another writer describes how a single giraffe drove away a hunting pair of lionesses and their eight cubs by kicking.[15] And there is one report of a female giraffe nearly decapitating a lion with a single front-on kick.[16]

—

Of course, the ape on two legs is the most fearsome predator of all. The two-legged ape is also a smart one, and giraffes have been hunted by humans ever since humans figured out how to kill a creature so much bigger, faster, and more alert than they. Mostly, this game was hunted for meat. The meat of adult males is "rank and uneatable by any one"[17]—or "entirely beastly."[18] But the meat from a female is said to pass the palate as satisfactorily as veal or venison,[19] or, to recall the words of the late-nineteenth-century British explorer, adventurer, and hunter Frederick Courtenay Selous: "There is no finer meat to be got in the whole world than that of a fat giraffe cow."[20]

A giraffe's ample bone marrow can seem "juicy and succulent,"[21] while the carcass may be stripped of its skin, sinews, and tail for a number of utilitarian purposes. The skin, unusually thick and tough, can be used to produce buckets and pots and drum heads, sandals, shields, and whips. A shield made from giraffe hide weighs less than one of rhino or buffalo hide yet will provide similar protection. Giraffe sinews, meanwhile, are unusually strong and historically were used for hunting bows, as strings for musical instruments, or for sewing, while the thick tail hairs were and sometimes still are valued as heavy thread to string beads on—for making amulets, bracelets, necklaces—and as a natural material for ceremonial fly whisks.[22]

Early North African hunters trapped giraffes using large, circular snares constructed, probably, from stiffly woven or knotted vegetable fiber and spread out on the ground.[23] Contemporary African hunters sometimes use comparable ground snares made from cable or cut sheet metal, and sometimes poachers will place wire neck snares into likely food trees, where the traps become nooses. Other traditional technologies for giraffe hunting include poisoned arrows, spears, camouflaged pitfalls, and fire to frighten and herd them off a cliff.[24]

But such traditional technologies pale in comparison with the horses and guns brought by Europeans. The Dutch-speaking Boers, who originally settled in southern Africa during the seventeenth century, hunted giraffes for fresh meat and the cured-meat product called *biltong*. They also hunted these animals for their commercially valuable hides, which provided the raw material for, among many other useful items, *sjamboks:* whips that, when made from a continuous strip of giraffe hide cut lip to heel, could reach more than six yards (six meters) in length. The Boers would run down entire herds of giraffes and slaughter them wholesale. These "innocent, graceful animals" had been, so a writer noted in an 1899 issue of *Scientific American*, "pot-hunted, shot down in droves, and destroyed in the greatest number possible in every direction. The extinction of this animal in South Africa is now threatened, and its preservation by legislation comes when it is almost too late."[25]

If the wasteful industriousness of the Boers was alarming, the puerile callousness of other white hunters was, or should have been, profoundly disturbing. This other kind of hunter killed not in response to the demands of hunger, as the Africans did, and not from ambition or industry, as the Boers did, but rather in the service of a personal kind of activity he liked to call "sport." Such a man hunted mainly to savor the emotions stimulated by manly exertion and risk, with the risk being astutely

In the early days of South African history the giraffe was the most abundant game in the Transvaal, Matabeland, and Orange Free State, but the creature has been killed off like our American buffalo, and the few remaining representatives of a noble race gradually driven north. For years past the giraffe has been a profitable quarry for the Boer hunters, and the animal was valued by them only because the hides were of commercial use. They were pot-hunted, shot down in droves, and destroyed in the greatest number possible in every direction. . . . On their hunting trips ten and fifteen years ago it was a common matter for one hunter to kill forty or fifty of these graceful animals in one day. The reason for this is that the giraffe is the most innocent of animals and easily hunted. They are absolutely defenceless, and there is hardly a case on record where a wounded giraffe turned upon the hunter. **—G.E.W., *SCIENTIFIC AMERICAN*, 1899**

At the report of the gun, and the sudden clattering of hoofs, away bounded the giraffes in grotesque confusion, clearing the ground by a succession of frog-like hops, and soon leaving me far in the rear. Twice were their towering forms concealed from view by a park of trees, which we entered almost at the same instant; and twice, on emerging from the labyrinth, did I perceive them tilting over an eminence immeasurably in advance. . . . In the course of five minutes, the fugitives arrived at a small river, the treacherous sands of which receiving their long legs, their flight was greatly retarded; and after floundering to the opposite side, and scrambling to the top of the bank, I perceived that their race was run. Patting the steaming neck of my good steed, I urged him again to his utmost, and instantly found myself by the side of the herd. The stately bull, being readily distinguishable from the rest by his dark chestnut robe, and superior stature, I applied the muzzle of my rifle behind his dappled shoulder with the right hand, and drew both triggers; but he still continued to shuffle along, and being afraid of losing him, should I dismount among the extensive mimosa groves with which the landscape was now obscured, I sat in my saddle, loading and firing behind the elbow, and then placing myself across his path, until, the tears trickling from his full brilliant eye, his proud form was prostrate in the dust. Never shall I forget the tingling excitement of that moment! Alone, in the wild wood, I hurraed with bursting exultation, and unsaddling my steed, sank exhausted beside the noble prize I had won. **—WILLIAM CORNWALLIS HARRIS, 1839**

calibrated. Rather than pursuing a sport, this kind of hunter was actually playing a game, a boy's game filled with the bold theatrics of masculinity.

Consider, for example, the case of Englishman William Cornwallis Harris. An engineer working for the British East India company, Harris was shipped off to South Africa in 1836 to recover from a fever acquired in India. South Africa stimulated another kind of fever. "From my boyhood upwards," Harris wrote in his 1839 hunting memoir, *Wild Sports of Southern Africa,* "I have been taxed by the facetious with *shooting madness,* and truly a most delightful mania I have found it."[26] In Africa, Harris was able to indulge this mania to the fullest, and when one day he discovered a herd of thirty-two giraffes, "industriously stretching their peacock necks to crop the tiny leaves which fluttered above their heads," he gave chase. At last the Englishman cornered the biggest giraffe of them all, a "stately bull" marked by his "dark chestnut robe, and superior stature." Harris sank seventeen bullets into the great animal before the giraffe began to falter. Finally, and with "tears trickling from his full brilliant eye," the animal's "proud form was prostrate in the dust." What a great event for the man who fired the bullets![27] Harris took enough time to make a drawing of the dead animal, then cut off the tail as a trophy, and left. Next day, he killed three more—and, having mastered the technique, continued to hunt giraffes after that. It was all good fun.

The red-bearded Scotsman Roualeyn George Gordon-Cumming killed more than a hundred elephants, as well as dozens of lions, hyenas, rhinos, hippos, giraffes—whatever crossed his fancy and gun sights simultaneously. The meat he gave away or left on the ground to rot. The skins and ivory he sent back to Britain in order to expand his own private collection of stuffed, glass-eyed simulacra of living nature.[28]

Inspired as a young boy by Gordon-Cumming's successful 1850 memoir (*Five Years of a Hunter's Life in the Far Interior of South Africa*), Frederick Courteney Selous grew up to become an explorer and big-game hunter in Africa. He considered hunting giraffes to be the sport "par excellence," in one year bagging eighteen of them, while often leaving their carcasses to rot.[29]

H. Anderson Bryden thought giraffes "beautiful and extraordinary creatures" who were "always a prime reward for the hunter's skill and labour"—even though "few beasts of the chase are more poorly endowed with means of defense."[30] Bryden once shot four in a quarter of an hour.

The American president Theodore Roosevelt left office in 1909 and proceeded directly to Africa, where he shot several giraffes as part of his determined slaughter of around five hundred big game animals.[31]

So it went. Many Europeans and Americans played the game: killing giraffes, sending the mounted heads home to hang in museums or on their walls, giving lectures to a fascinated public, writing self-serving memoirs extolling the pleasures to be found in spilling blood and acting out the manly role on an African stage. In truth, however, killing giraffes was not very difficult. As one journalist declared in an 1897 article for *Harper's Weekly,* "The pleasure of a giraffe-hunt is indisputable." It is also a short-lived pleasure, he went on to insist, one that "no real sportsman will repeat more than twice." The real excitement was in the chase itself, but with a good horse and the right gun and bullet, killing a giraffe lacked that critical element of the hunter's standard game: the genuine sense of danger, of risk, since "no animal could be more gentle or defenceless."[32]

The following photographs consider a giraffe's relationship with "others" of various sorts. First come others of the same species: giraffes racing in a coherent herd. Second are the others involved in a mutually beneficial partnership: tickbirds. One gets the impression that tickbirds are an occasional distraction or nuisance, as in the photograph of a giraffe who appears distracted by his fluttering companion; but the birds do keep a steady watch for ticks, as the next photo suggests.

Third are the occasionally helpful, or at least not harmful, others of various mammal species. Elephants and giraffes, it seems to me, have little to gain from one another. The giraffes watching wild donkeys and a migrating line of wildebeests seem, at first, merely curious. At the same time, though, standing in a group with other mammals who fear the same predators should be calming and, if the others are also sufficiently alert, useful.

Fourth are the predatory others, with the photographs featuring the most fearsome nonhuman predator of all: lions. The mother giraffe apparently does what she can to fend off the lions, but she is ultimately helpless against two young females backed by an opportunistic male. There is a chilling beauty to lions, and of course nature has given them, too, an important role in the drama of existence.

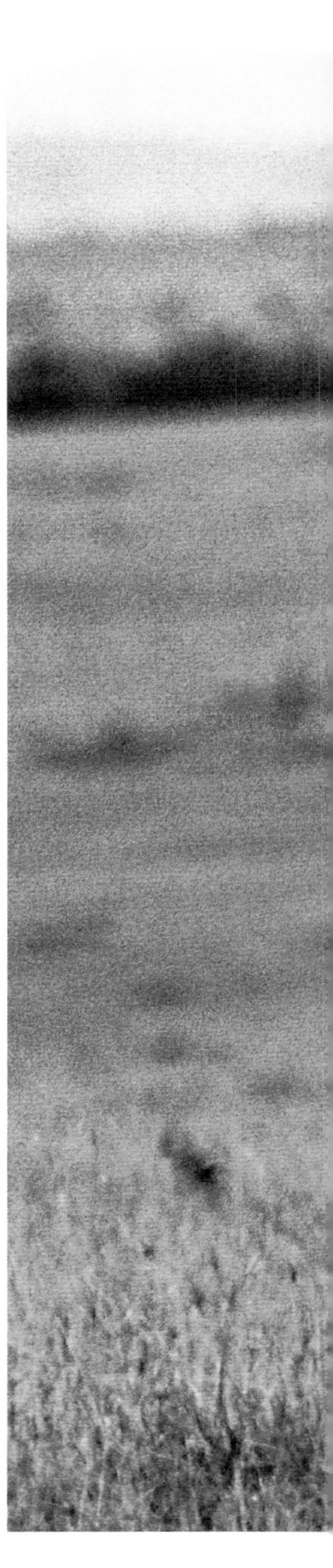

KINDS

BY THE START OF THE NINETEENTH CENTURY, European scientists had been introduced, more or less, to giraffes from both the northern and southern ends of Africa. Karl Linnaeus based his 1735 classification of giraffes on some crude sketches and rough reports of captive northern giraffes seen by a few travelers to the Middle East. By the 1790s, François Le Vaillant's memoir of travels in southern Africa, along with his shipments of specimen hides and bones, introduced southern giraffes in a more material fashion to European experts. But were the northern and southern giraffes the same kind of beast?

French zoologist Étienne Geoffroy Saint-Hilaire, having at the start of the nineteenth century read the memoirs and studied the specimen materials Le Vaillant had shipped to Paris, decided there was only one kind of giraffe. But then along came the zarafa. Saint-Hilaire, given the responsibility and honor of leading that animal in a 550-mile cross-country promenade from Marseille to Paris during the spring of 1827, eventually concluded that the pasha's zarafa and Le Vaillant's giraffe were of distinctly different kinds. Different enough, he thought, to be individuals from two different species.

The debate continued. At the British Natural History Museum, Sir Richard Owen examined a small collection of skull specimens sent from northern and southern Africa, and agreed in 1841 that giraffes really should be divided into two species.

Writing from the Swedish Museum of Natural History, zoologist Carl Jakob Sundevall begged to disagree. There was only one species, he declared, and the northern and southern types were mere "races" or subspecies, mainly distinguishable by the length of their hair. From Germany, Carl Hagenbeck, an active dealer in exotic animals and the founder of a successful private zoo, experimentally kept northern giraffes in unheated stables during a cold European winter and demonstrated that, in response to the seasonal challenge, their hair grew to be two to three times longer than normal. Hair length, then, was a quick and natural response to weather or climate, meaning that the feature was superficial—insufficient to mark a difference in species.[1]

The nineteenth-century flood of Europeans into Africa, many of them handy with a gun, produced a countercurrent of animal specimens flowing into Europe, and as more giraffes were seen and shot in more and more different places in Africa, the discovery of giraffes proceeded into a discovery of giraffe diversity.

In 1893, for example, a very odd giraffe indeed was shot near Lake Rudolph (now Lake Turkana) in East Africa's Great Rift Valley. The hide taken from this animal was strikingly different, thought taxonomist O. Thomas, examining it carefully in the British Natural History Museum, from any he had seen before, marked as it was by a cream-colored netting over liver-colored polygonal plates. It was, Thomas decided, *reticulated,* to use a zoological term that referred to the clear net or network quality of the pattern. The specimen was still from a member of the single species of giraffes found in Africa, he believed, but surely one of two distinctive subspecies, which he identified as the *reticulated* and *blotched* subspecies.[2]

In 1899, William Edward de Winton looked over some skulls in the British Natural History Museum and decided it was most appropriate to think of giraffes as two different species, northern and southern, based on their horn patterns. Northern giraffes, at least the males, de Winton believed, had three distinct horns, while a typical male of the southern species only had two. He identified these two species as *Giraffa camelopardalis* and *Giraffa capensis*—prematurely, as it turned out, since the still expanding collection of skull specimens would soon show an unpredictable variation in the number of horns for males across the continent.

Finally, in 1904, the British Natural History Museum's Richard Lydekker methodically reviewed all the literature and specimen materials available in Europe at the time. He pronounced that giraffes were indeed two species, but the distinction was not northern versus southern. It was reticulated versus blotched, two species he called *Giraffa reticulata* and *Giraffa camelopardalis*. The blotched species, however, was so widely distributed and diverse that it should be divided into a total of ten subspecies.[3]

How many kinds of giraffes? How many species? How many subspecies? The debates may seem confusing and academic, irrelevant to anyone living outside the dusty confines of a museum, but they reflect the real-world truth of an unusual diversity among giraffes. In any case, by the middle of the twentieth century, most experts had simplified the discussion a little by settling into a consensus that giraffes are a single species, *Giraffa camelopardalis,* with nine subspecies, the subspecies being mainly distinguishable by coat coloring and pattern combined with geographical distribution.

—

When those nineteenth- and early twentieth-century zoologists in London, Paris, Stockholm, and elsewhere were debating the species structure of giraffes—one species or two? two subspecies or ten?—they were mainly comparing physical appearance by looking at some collections of bones and hides. They had trouble agreeing in part because, with only a small number of decent specimens on hand, it was very difficult to see beyond the normal variation between individuals and appreciate the average variation distinguishing one group from another. And even as the specimen samples expanded in the twentieth century, the learned fellows still had trouble agreeing because they had no precise way to distinguish species from subspecies.

A species is a collection of individuals who generally look alike. Indeed, we might consider the species *type* as a hypothetical average of all individual features for all members of the species. But for most people, the distinction between one species and another requires no theory or study or arithmetical assessment. It arrives intuitively. Individuals within a species generally look alike. Individuals outside the species generally look different. In most ordinary circumstances, intuition is a fair guide for distinguishing species. Dog is not cat. But the visible similarity of individuals within a species expresses an underlying truth that those individuals are part of the same gene pool. A species, then, might be described as a group of individuals—whether a hundred or seven billion—who are steadily interchanging genetic material.

A species will change over time, of course, but it can also change over space. That is to say, a species can divide. The usual reason for such a division would be a physical event. Part of the group crosses that big, raging river, and the other part stays behind. Pressured by differing ecological circumstances, the two groups gradually diverge in their average genetic makeup. This genetic division begins to produce an overt, physical one that may be minor at first, but given, say, ten thousand or a hundred thousand years, can become major.

Subspecies develop this way. An environmental barrier splits a population in two, and the two groups stop exchanging genetic material. Given a certain amount of time, the two groups drift apart genetically and therefore in appearance. The rather subtle marks for a subspecies arrive earlier in the process of drifting apart. The stronger differences between species emerge later. Time is one way to think about this, but the critical difference is that subspeciation should be reversible, while speciation should not. Reversible: If two subspecies are physically rejoined, they start to interbreed and so return to an open exchange of genetic material. Irreversible: If two species are physically rejoined, they will still not readily interbreed, either because they choose not to or because they are incapable of doing so. In any case, the two species will never again exchange genetic material.

Techniques for genetic analysis that have been perfected in the past decade have enabled us to follow the story of giraffe speciation to a degree never before possible. We know that ancestral giraffes arrived in northern Africa several million years ago, then over the next few million years spread broadly across the continent, north to south, east to west. Genetic studies now show that the division of giraffes into separate groups happened as the consequence of intense climatic change that created various habitat barriers between 1.62 million and 13,000 years ago.[4]

The same genetic studies also indicate that the giraffes of Africa divide into a minimum of six "largely independent gene pools" that ought to be recognized as six species, not subspecies. (Not all populations were sampled, which is the reason for considering six a minimum.) The argument for this species-level division combines a highly technical assessment with some behavioral analysis. Giraffes are very large and mobile. They live in loosely constructed social groups, and sometimes in some places (populations vary markedly in ecology and range) they will roam over hundreds of square miles. These features predict that the genetic differences between adjacent giraffe populations should be comparatively low, since "the extent of gene flow is related to the dispersal potential of individuals."[5] This principle holds true for such other large and mobile African mammals as elephants and buffalos, who ordinarily do not show a significant genetic differentiation between adjacent populations. Giraffes, however, do.

The point becomes particularly clear for three East African groups—the Rothschild's, reticulated, and Masai giraffes—who happen to live in adjacent and often continuous acacia scrub and woodland habitat. The three groups are not separated by any obvious habitat or topographic barriers, and yet genetic studies show they are not mating with each other. The experience of zoos indicates that different kinds of giraffes are capable, in highly artificial circumstances, of mating to create hybrid offspring. Yet the genetic studies demonstrate that wild giraffes are generally not doing so, even when the environment would seem to make it easy. Why? If the wild giraffes are not limited by environmental or physical barriers, then perhaps they are limiting themselves by choice.

This idea implies in turn that a Masai giraffe looking at, say, a Rothschild's giraffe sees not another giraffe but rather an animal of another species, a distinction possibly based on the visual recognition of coat patterns. It also indicates that scientists looking at Masai and Rothschild's giraffes should be thinking of two species, rather than two subspecies, since those two populations are not exchanging genetic material, even in the absence of obvious habitat barriers.

For conservationists, this new way of thinking about giraffes makes the task of protecting them even more urgent. The total number of giraffes in the world is estimated at fewer than 75,000 individuals, and they are increasingly scattered in small pockets of declining habitat across the continent.[6] Like all other large animals, including the other iconic giants—elephants, rhinos, lions—they are disappearing as a result of human exploitation and encroachment. Human memory tells the same story: At the Samburu National Reserve, Karl and I spoke with a young Samburu warrior named Julius Lesori, who as a boy herding cattle twenty years ago in the remote Wamba area would see giraffes every day. Now he works as a guide in the Samburu Reserve, while his three children herd the cattle in Wamba—without ever seeing a giraffe. The giraffes have been hunted out. If his children want to see giraffes, Lesori declared, they have to visit the reserve.

Seventy-five thousand is an alarmingly small number, equivalent to the number of new human beings added to our planet every few hours. But the reality of the situation becomes a good deal more distressing as we divide such a small number in six or more ways to identify six or more highly threatened or endangered species: some with merely several thousand individuals left, some with at best a few hundred.[7]

—

From a distance, giraffes can seem improbably slow, quiet, and sometimes strangely curious. Most of what they do takes place very slowly, in an underwater kind of motion. They are quiet enough that earlier observers mistakenly described them as mute, physiologically incapable of producing sound. They are characteristically quiet, yes, but not constitutionally silent.

They are disarming in that quietude; and, in their appearance, sometimes, they can be astonishing. When walking, they move both legs on one side at a time. Stretched beyond reason along the vertical axis, they commune with birds. Their camel-like faces are topped with skin-covered horns shaped like chimney pots or little smokestacks. Dreamily, they slip into the trees, pass into the horizon, fade away into the night. Yet standing tall on the noonday horizon, horns up and ears out, they show the bold profile of a four-legged communications tower with four antennae on top.

At a distance, they can appear spindly and even fragile, like nature's archetype of the ballet dancer, but for all their grace and elegance, these are very big, very powerful creatures. You will be surprised by their mass. Up close, you take note of those giant cloven hooves, split black spades that, propelled by a long-legged lever, can damage a lion. Even their necks, so narrow and graceful at a distance, look, up close, lean and powerful. Dangerous, even. Their heads are bony, with giant lash-covered eyes that stare at you darkly, and you recognize beneath the skin a skull the size and heft of a blacksmith's anvil. A solid blow would knock you down and out. And yet, if you are lucky, you might be touched by a long, dark tongue as gentle and rough as a cat's.

You can study them up close, but they just might study you back. Sometimes they will do it with a straight-on look, sometimes with an over-the-shoulder look or a fleeting sideways glance. Their necks, supple and snakelike, enable a gathering of multiple perspectives. A neck will flick a ribbony twist, allowing the eye to glance first from one side and then the other, and then, as the body turns away, will flick another twist so that yet another gaze is registered from over the shoulder. That peering at you one way and then another seems to me a lovely sort of subtle, sidelong observation: as if you are the apparition, you the oddity, you the wispy tendril of smoke or the rising spirit. But what a tragic loss for ourselves and our children and grandchildren, as in reality these ethereal creatures decline and dissolve and drift away, bit by bit, kind by kind, from this our fragile world.

The opening image shows a puzzle of Masai giraffes—one of the "blotched" species, according to a system proposed by British taxonomist O. Thomas in the late nineteenth century. That giraffes can identify and will prefer the markings of their own group makes good sense, and this photo of Masai giraffes clustered tightly together seems to embody the concept. Such a preference might account for the division of giraffes into at least six clearly defined genetic groups recently described by some geneticists.

There are a few bright spots in the current state of giraffes in Africa—notably, perhaps, in the work being done to protect giraffes in Namibia, where a low level of human population combined with political will and a cultural interest in giraffes may be making a difference. But the final truth is that giraffes are being displaced rapidly across most parts of their former range by an unprecedented human expansion. I have chosen these closing photographs to evoke the giraffe farewell, a recessional march into the night of nonexistence.

ACKNOWLEDGMENTS

We are especially grateful to three top experts who generously advised us on parts of this book: Anne Innes Dagg, Richard D. Estes, and Elizabeth Marshall Thomas. A fourth expert, Julian Fennessy, member of the board of trustees of the Giraffe Conservation Foundation and chair of the International Giraffe Working Group (IUCN / SSC ASG), not only served as our primary scientific advisor but also toured the deserts of northwestern Namibia with us and helped us find desert-adapted giraffes. To him we are doubly, or triply, grateful. Thanks also to the Giraffe Conservation Foundation.

In the Democratic Republic of the Congo, we were helped in material ways by J. J. Marilanga wa Tsaramu and Michele Moyakeso at the Okapi Faunal Reserve in the Ituri Forest. Others in the DR Congo who graciously assisted were Pierre Bato, Abeli Doki, Edmond Malemo, John-Prince M'Bayaa, Bernard Mtongani, and Mzee Myanamenge. We are also grateful for special information provided by Julius Lesori at the Samburu National Reserve in Kenya and Thekla Tsaraes at Twyfelfontein in Namibia.

Closer to home, we thank Blaine Gaustad for expert advice on Chinese history; Wyn Kelley for an expert general review of the manuscript; and Sheila Levine at University of California Press, who originally encouraged the idea for this book and made possible the first transformation from idea to image and word. We are also deeply indebted to those at University of California Press who made possible the second transformation, from image and word to informative and lasting thing of beauty. Thank you Dore Brown, Blake Edgar, Nicole Hayward, Lynn Meinhardt, and Jan Spauschus.

NOTES

Spirits

1 Pager n.d., p. 7.
2 Lewis-Williams 1983, p. 11.
3 Quoted in Lewis-Williams 1983, p. 13.
4 Barrow 1802, p. 226.
5 This argument is developed by Lewis-Williams (1983). A fuller exploration of shamanism and neolithic religions from around the world can be found in Lewis-Williams 2002 and Lewis-Williams and Pearce 2005.
6 Bleek, quoted in Watson 2002, p. 77. All the Bushman languages include various clicks as standard phonemes. The dental click (represented as /) sounds like the *tsk* in the exclamation *tsk tsk tsk*. The alveolar click (≠) is similar to the dental click except the sound is created farther back in the mouth. The lateral click (//) resembles the *cluck* of tongue against palate that a person might use to encourage movement in a reluctant horse. The alveol-palatal click (!) is produced by raising the tongue to the palate and dropping it crisply enough to produce a pop. See Thomas 2006, p. xiv.
7 According to Lewis-Williams (1983), the /Xam poisons were manufactured from various plants and snake venom. The anthropologist Elizabeth Marshall Thomas informed me (personal communication) that the Bushman poison she knew came from the grubs of *Diamphidia simplex* beetles living in marula trees:

> Nothing much grew under the trees, so why did the people dig there? But they did. The grubs were poison in the pupa state, made pupa casings about 18 inches down in the sand that were very hard to see. The people would open the casing and extract the grub, then mush it up, take off the head, and spread the insides on an arrow. About six would poison an arrow. The poison works on hemoglobin, so it must be injected into the blood stream and takes a long time, but there is no antidote. One drop will kill a person in about a day, a kudu in three or four days, during which they have to track the animal. I feel pretty sure that all Bushmen would use this poison (if they could get it), because as far as I know all groups of Bushmen used small arrows, which are dart deliverers—no hope of a Bushman arrow killing anything right away. Also the arrows were designed to fall apart after striking, so a kudu coudn't grab it with his teeth and pull it out.

8 Thomas 2006. Also Thomas (1958) 1989, pp. 127–38.
9 Thomas 2006, pp. 104, 105.
10 Thomas 2006, p. 92.

Chimeras

1 According to Rice 1983, p. 5.
2 Based on E. E. Rice of Oxford's translation of Athenaeus of Naucratis's quotation of Kallinexos of Rhodes's account: Rice 1983, pp. 7–25.
3 Dawson 1927, p. 482.
4 De Villiers and Hirtle 2002, p. 9.
5 De Villiers and Hirtle 2002, p. 46.
6 Dawson 1927; Spinage 1968, pp. 33, 34.
7 Spinage 1968, pp. 34, 35.
8 Kistler 2006.
9 According to Belozerskaya 2006.
10 Strabo 1930, pp. 17, 19.
11 According to Strabo 1930, p. 119.
12 Hubbell 1935, p. 73.
13 Diodorus quoted in Hubbell 1935, p. 74.
14 Quoted in Spinage 1968, p. 41.
15 Strabo 1930, p. 337.
16 Belozerskaya 2006, pp. 75, 76.
17 Schiff 2011, p. 98.
18 Quoted in Laufer 1928, p. 60.
19 Laufer 1928, p. 58.
20 Heliodorus 1957, p. 265.

Unicorns

1 Duyvendak 1949, pp. 13–15; also Laufer 1928, pp. 41–54.
2 Duyvendak 1949, p. 32.
3 Duyvendak 1949, p. 32.
4 Williams (1931) 1976, pp. 413–15; Wolfram 1986, pp. 302–4.
5 Chan 1988, pp. 214–18.
6 Atwell 2002, p. 86.
7 Ebrey 1999, pp. 192, 193; Fairbank and Goldman 2006, pp. 129, 130.
8 Chan 1988, pp. 212–14; also Langlois 1988.
9 Chan 1988, pp. 218–21.
10 Tsai 1996, pp. 156, 157.
11 Chan 1988, pp. 232–36.
12 Duyvendak 1949, p. 33.
13 Duyvendak 1949, p. 33.
14 Duyvendak 1949, pp. 33, 34.
15 Duyvendak 1949, p. 34.

Zarafas

1 Spinage 1968, p. 53.
2 Spinage 1968, p. 52.
3 Spinage 1968, pp. 51, 52.
4 White 1954, pp. 230–37.
5 White 1954, pp. 43, 134.
6 White 1954, p. 33.
7 Wikipedia contributors, "Frederick II, Holy Roman Emperor," *Wikipedia, The Free Encyclopedia*, http://en.wikipedia.org/wiki/Frederick_II,_Holy_Roman_Emperor, 2012.
8 Wikipedia contributors, "Frederick II."
9 Spinage 1968, p. 56.
10 Laufer 1928, p. 71.
11 Belozerskaya 2006, p. 121.
12 Belozerskaya 2006, p. 90.
13 Laufer 1928, pp. 79, 80.
14 Belozerskaya 2006, p. 110.
15 Quoted in Spinage 1968, p. 73.
16 Quoted in Spinage 1968, p. 74.

17 Belozerskaya 2006, p. 128.
18 Quoted in Laufer 1928, p. 74.
19 Quoted in Laufer 1928, p. 75.
20 Spinage 1968, p. 60.
21 Spinage 1968, p. 68.
22 Laufer 1928, pp. 75, 76.
23 Belon 1553, p. 118; English translation in Laufer 1928, pp. 84, 85.
24 Topsell (1607) 1658, p. 79.
25 Topsell (1607) 1658, p. 79.
26 Spinage 1968, pp. 79, 80.
27 Johnson 1755, p. 318.
28 Quoted in Spinage 1968, p. 81.
29 Quoted in Spinage 1968, p. 82.
30 Spinage 1968, p. 85.
31 Allin 1998, pp. 47, 48.
32 Allin 1998, pp. 176, 177.

Giraffids

1 Stiles 2010, p. 47.
2 Stanley 1890, p. 490.
3 Quoted in Spinage 1968, p. 156; see also Johnston 1923, p. 280.
4 Material on okapis, including the story of their discovery, is based on Lankester 1903, pp. 279–83; Spinage 1968, pp. 143–61; and Johnston 1923.
5 Dagg and Foster (1976) 1982, p. 173.
6 Spinage 1968, pp. 103–5.
7 Mitchell and Skinner 2003, p. 53.
8 Spinage 1968, p. 114.
9 Mitchell and Skinner 2003, p. 57.
10 Dagg and Foster (1976) 1982, p. 59.
11 Mitchell and Skinner 2003, p. 58.
12 Mitchell and Skinner 2003, p. 61.

Bodies

1 Mossop 1947, pp. 33–35, 49; see also Spinage 1968, p. 87.
2 Le Vaillant 1796, p. 173.
3 Le Vaillant 1796, p. 262.
4 Le Vaillant 1796, p. 268.
5 Owen 1839; see also Guggisberg 1969, pp. 17–20.
6 Dagg and Foster (1976) 1982, p. 71.
7 Dagg and Foster (1976) 1982, p. 72.
8 Dagg and Foster (1976) 1982, p. 72.
9 Mitchell and Skinner 2003, p. 66; Van Sittert, Skinner, and Mitchell 2010, p. 477.
10 Mitchell and Skinner 2003, p. 65.
11 Dagg and Foster (1976) 1982, p. 37.
12 Dagg and Foster (1976) 1982, p. 97; see also Arbuthnot 1954.
13 Dagg and Foster (1976) 1982, p. 65.
14 Du Toit 1990.
15 Mitchell and Skinner 2003, p. 69.
16 Van Sittert, Skinner, and Mitchell 2010.
17 Mitchell and Skinner 2003, p. 68.
18 Spinage 1968, p. 103.
19 Cited in Dagg and Foster (1976) 1982, p. 173, referring to Dagg 1965.
20 Dagg and Foster (1976) 1982, pp. 172, 173.
21 The same question could be asked about giraffe horns: males use them to fight each other but never employ them for defense against predators; females do not seem to use them at all. In fact, across the spectrum of African hoofed mammals, we find females sporting but virtually never using male weaponry, a seemingly odd fact that may be explained as defense (of self and young) from males of the species through male mimicry. As Estes 1991 (p. 403), summarizes: Among bovids "females mimic male secondary sexual characters to buffer their male offspring against the aggression of dominant males. . . . Horns, and other secondary characters that provoke despotic competition as male sex symbols, become ambiguous and no longer release aggression when females possess similar ornaments."
22 Dagg 1962.
23 The walking gait is unusual but not unique; according to Dagg, okapis have a similar gait. See Dagg 1960.
24 Dagg and Foster (1976) 1982, pp. 93–97.
25 http://wiki.answers.com/Q/How_long_does_a_giraffe_sleep_in_a_day; see also Dagg and Foster (1976) 1982, pp. 89, 90.
26 Mitchell and Skinner 2003, p. 68.
27 Warren 1974; Dagg and Foster (1976) 1982, p. 169.
28 Warren 1974.

Behaviors

1 Dagg 2006, p. 2.
2 Dagg 2006, p. 5.
3 Innis 1958, pp. 245–49.
4 Dagg 2006, pp. 88, 89.
5 Dagg 2006, p. 89.
6 Dagg 2006, p. 34.
7 Dagg 2006, p. 69.
8 Dagg and Foster (1976) 1982, p. 74.
9 Dagg 2006, p. 102; Innis 1958, pp. 249–53.
10 Over a hundred: Julian Fennessy, personal communication.
11 Dagg and Foster (1976) 1982, p. 131.
12 Spinage 1968, p. 130.
13 Dagg 2006, p. 118.
14 Coe 1967 provides a useful analysis of necking.
15 Dagg 2006, p. 74.
16 Innis 1958, pp. 259, 260.
17 Dagg 2006, pp. 73, 74; Dagg 1983.
18 Spinage 1968, p. 130.
19 Dagg 2006, p. 69.

Mothers

1 Dagg 2006, p. 81.
2 Innis 1958, p. 263, refers to her initial impression of "a rather weak maternal instinct."
3 Foster 1966, p. 143.
4 Dagg and Foster (1976) 1982, p. 132.
5 Stephan 1925, p. 61.
6 Savoy 1966, p. 202.
7 Dagg and Foster (1976) 1982, p. 136.
8 Premature births are likely to be a problem when the baby is too short to nurse in the normal standing position, as Steinemann 1955 (p. 40) reported for the Basel Zoo.
9 Pournelle 1955.
10 Hediger 1955 (pp. 91, 92) describes a mother as "clearly afraid" of her underweight and "far too small" infant; yet the mother slowly became used to the infant and showed signs of a growing attachment. A "keener interest gradually awakens," as the notes have it, but, unfortunately, the mother accidentally moved too close

and stepped on the calf's leg, breaking it. The infant was removed.

11 Dagg and Foster (1976) 1982, p. 132.

12 "Giraffe Midwives" 1965.

13 As reported by Moss (1975) 1982, p. 51.

14 Moss (1975) 1982, p. 51.

15 Backhaus 1961.

16 Paraphrase from Moss (1975) 1982, p. 52.

17 Reported in Moss (1975) 1982, p. 53.

18 Langman 1982, p. 95.

19 Langman 1982, p. 97.

20 Fennessy 2004, pp. 199–201.

21 Langman 1982, p. 103.

22 Muller 2010, pp. 20–23.

23 Muller 2010, p. 20.

24 Muller 2010, p. 21.

25 Muller 2010, p. 22.

26 As Muller 2010 notes, this behaviorial complex, the reaction to a dead conspecific, is worth comparing to the several reports of elephants responding to the death of other elephants as if in grief. See, for example, Douglas-Hamilton et al. 2006; Joshi 2010; and McComb, Baker, and Moss 2006.

Others

1 See Estes 1974, for useful comments on social group structure among African bovids.

2 Researcher Barbara Leuthold's mean group size at Tsavo East National Park in Kenya was based on including "groups" of one individual in her sample; it is not clear if another study she cites, done in Tanzania's Serengeti National Park and discovering a mean group size of 9.16, likewise included solitary individuals in its sample. See Leuthold 1979.

3 Leuthold 1979, p. 25.

4 Leuthold 1979, p. 27.

5 Van der Jeugd and Prins 2000; Pratt and Anderson 1982.

6 Quoted in Milius 2003.

7 Silk 2002. Julian Fennessy (personal communication): "Such questions are still being asked in large part because no one has ever conducted a long-term study of a wild giraffe population and its individuals—which limits our understanding compared to what is known about, say, elephant sociality, based on the studies of the Amboseli Trust, or about chimpanzees as a result of work in places such as Jane Goodall's site in Gombe Stream National Park, Tanzania." We still have much to learn about giraffes.

8 Milius 2003.

9 Tarou, Bashaw, and Maple 2000.

10 Bashaw et al. 2007, p. 48 (table 1).

11 Pratt and Anderson 1985.

12 Bashaw et al. 2007, p. 50.

13 Other bird species occasionally do the same job as tickbirds—for example, starlings in northwest Namibia.

14 Kingdon 1997; Dagg 1971.

15 Roberts 1969.

16 Campbell 1952, p. 80.

17 Brooks 1987, p. 983.

18 Puxley 1929, p. 61.

19 Dagg and Foster (1976) 1982, p. 10.

20 Selous 1908, p. 217.

21 Brooks 1987, p. 983.

22 Dagg and Foster (1976) 1982, pp. 8, 9.

23 Spinage 1968, p. 25.

24 Dagg and Foster (1976) 1982, pp. 8, 9.

25 G. E. W. 1899, p. 259.

26 Harris 1839, p. xvii.

27 Harris (1840/41) 1969, pp. 236, 237.

28 Gordon-Cumming and George 1850.

29 According to Dagg and Foster (1976) 1982, p. 10; see Selous 1907 and Selous 1908.

30 Bryden 1893, pp. 304, 322, 330.

31 Roosevelt 1910, pp. 117–25.

32 Brooks 1987, p. 983.

Kinds

1 Dagg and Foster (1976) 1982, p. 49; Street 1956.

2 Thomas 1894.

3 Dagg and Foster (1976) 1982, pp. 46–54.

4 Brown et al. 2007.

5 Brown et al. 2007.

6 Julian Fennessy (personal communication).

7 Fennessy 2007.

REFERENCES

Allin, Michael. 1998. *Zarafa: A Giraffe's True Story from Deep in Africa to the Heart of Paris.* New York: Delta.

Arbuthnot, Thomas S. 1954. *African Hunt.* New York: W. W. Norton.

Atwell, William S. 2002. "Time, Money and the Weather: Ming China and the 'Great Depression' of the Mid-Fifteenth Century." *Journal of Asian Studies* 61 (1): 83–113.

Backhaus, Dieter. 1961. *Beobachtungen an Giraffen in zoologischen Gärten und frier Wildbahn.* Brussels: Institut des Parcs Nationaux du Congo et du Ruanda-Urundi.

Barrow, John. 1802. *An Account of Travels into the Interior of Southern Africa in the Years 1797 and 1798.* New York: G. F. Hopkins.

Bashaw, Meredith J., Mollie A. Bloomsmith, Terry L. Maple, and Fred B. Bercovitch. 2007. "The Structure of Social Relationships among Captive Female Giraffe (*Giraffa camelopardalis*)." *Journal of Comparative Psychology* 121 (1): 46–53.

Belon, Pierre. 1553. *Les Observations de plusieurs singularitez et choses mémorables.* Vol. 2. Paris: G. Corrozet.

Belozerskaya, Marina. 2006. *The Medici Giraffe: And Other Tales of Exotic Animals and Power.* Boston: Little, Brown.

Bercovitch, Fred B., and Philip S. M. Berry. 2009. "Ecological Determinants of Herd Size in the Thornicroft's Giraffe of Zambia." *African Journal of Ecology* 48 (4): 962–71.

Brooks, Sydney. 1897. "Giraffe Hunting." *Harper's Weekly.* October 2, 983.

Brown, David M., Rick A. Brenneman, Klaus-Peter Koepfli, John P. Pollinger, Borja Mila, Nicholas J. Georgiadis, Edward E. Louis, Gregory F. Grether, David K. Jacobs, and Robert K. Wayne. 2007. "Extensive Population Genetic Structure in the Giraffe." *BMC Biology* 5 (57): 1–13.

Bryden, H. Anderson. 1893. *Gun and Camera in South Africa.* London: Edward Stanford.

Campbell, Roy. 1952. *Light on a Dark Horse.* Chicago: Henry Regnery.

Chan, Hok-Lam. 1988. "The Chien-wen, Yung-lo, Hung-hsi, and Hsüan-te Reigns, 1399–1435." In *The Cambridge History of China.* Vol. 7, *The Ming Dynasty, 1368–1644, Part I,* ed. Frederick W. Mote and Denis Twitchett, 182–304. Cambridge: Cambridge University Press.

Coe, Malcolm J. 1967. "'Necking' Behaviour in Giraffe." *Journal of Zoology, London* 151 (3): 313–21.

Dagg, Anne Innis. 1960. "Gaits of the Giraffe and Okapi." *Journal of Mammalogy* 41 (2 [May]): 282.

———. 1962. "The Role of the Neck in the Movements of the Giraffe." *Journal of Mammalogy* 43 (1 [Feb.]): 88–97.

———. 1971. "*Giraffa camelopardalis.*" *Mammal Species* 971 (5): 1–8.

———. 1983. "Homosexual Behavior and Female-Male Mounting in Mammals—A First Survey." *Mammal Review* 14: 155–85.

———. 2006. *Pursuing Giraffe: A 1950s Adventure.* Waterloo, Ontario: Wilfrid Laurier University Press.

Dagg, Anne Innis, and J. Bristol Foster. (1976) 1982. *The Giraffe: Its Biology, Behavior, and Ecology.* Malabar, FL: Krieger.

Dawson, Warren R. 1927. "The Earliest Records of the Giraffe." *The Annals and Magazine of Natural History* (9): 478–85.

Descartes, Réné. (1637) 1993. "Discourse on Method." In *Environmental Ethics: Divergence and Convergence,* ed. S. J. Armstrong and R. G. Botzler, 281–85. New York: McGraw-Hill.

De Villiers, Marq, and Sheila Hirtle. 2002. *Sahara: A Natural History.* New York: Walker.

Douglas-Hamilton, Iain, Shivani Bhalla, George Wittemyer, and Fritz Vollrath. 2006. "Behavioural Reactions of Elephants towards a Dying and Deceased Matriarch." *Applied Animal Behaviour Science* 100 (1–2): 87–102.

Du Toit, J. T. 1990. "Feeding-height Stratification among African Browsing Ruminants." *African Journal of Ecology,* no. 28: 55–61.

Duyvendak, J. J. L. 1949. *China's Discovery of Africa.* London: Arthur Probsthain.

Eberhard, Wolfram. (1983) 1986. *A Dictionary of Chinese Symbols.* London: Routledge & Kegan Paul.

Ebrey, Patricia B. 1999. *The Cambridge Illustrated History of China.* Cambridge: Cambridge University Press.

Estes, Richard D. 1974. "Social Organization of the African Bovidae." In *The Behaviour of Ungulates and Its Relation to Management,* ed. V. Geist and F. Walther, 166–205. Morges, Switzerland: International Union for the Conservation of Nature and Natural Resources.

———. 1991. "The Significance of Horns and Other Male Secondary Sexual Characters in Female Bovids." *Applied Animal Behaviour Science* 29: 403–51.

Fairbank, John K., and Merle Goldman. 2006. *China: A New History.* 2nd enlarged edition. Cambridge, MA: Belknap Press.

Fennessy, Julian. 2004. "Foraging Ecology." In "Ecology of Desert-Dwelling Giraffe *(Giraffa camelopardalis angolensis)* in Northwestern Namibia," 160–215. PhD diss., University of Sydney.

———. 2007. "GiD: Development of the Giraffe Database and Species Survival Report." *Giraffa* 1 (2): 2.

———. 2008. "An Overview of Giraffe (*Giraffa camelopardalis*) Taxonomy, Distribution and Conservation Status, with a Namibian Comparative and Focus on the Kunene Region." *Journal NWG / Journal NSS* 56: 1–16.

Foster, J. B. 1966. "The Giraffe of Nairobi National Park: Home Range, Sex Ratios, the Herd, and Food." *East African Wildlife Journal* 4: 139–48.
G. E. W. 1899. "The Boers and the Giraffe." *Scientific American* 81 (Oct. 21): 259.
"Giraffe Midwives at a Birth." 1965. *African Wild Life*, no. 19: 323.
Gordon-Cumming, Roualeyn George. 1850. *Five Years of a Hunter's Life in the Far Interior of South Africa*. 2 vols. New York: Harper & Brothers.
Guggisberg, C. A. W. 1969. *Giraffes*. New York: Golden Press.
Harris, William Cornwallis. 1839. *Wild Sports of Southern Africa*. London: John Murray.
———. (1840/41) 1969. *Portraits of the Game and Wild Animals of Southern Africa*. Cape Town: A. A. Balkema.
Hediger, Heini. 1955. *Studies of the Psychology and Behaviour of Captive Animals in Zoos and Circuses*. New York: Criterion.
Heliodorus. 1957. *An Ethiopian Romance*. Trans. and intro. Moses Hadas. Philadelphia: University of Pennsylvania Press.
Hubbell, Harry M. 1935. "Ptolemy's Zoo." *Classical Journal* 31 (2): 68–76.
Innis, Anne. 1958. "The Behaviour of the Giraffe, *Giraffa camelopardalis*, in the Eastern Transvaal." *Proceedings of the Zoological Society of London* 131 (2): 245–78.
Johnson, Samuel. 1755. *A Dictionary of the English Language*. London: Printed by W. Strahan for J. and P. Knapton; T. and T. Longman; C. Hitch and L. Hawes; A. Millar; and R. and J. Dodsley.
Johnston, Harry H. 1923. *The Story of My Life*. Garden City, NY: Garden City.
Joshi, R. 2010. "How Social Are Asian Elephants, *Elephas maximas*?" *New York Science Journal* 3 (1): 27–31.
Kingdon, Jonathan. 1997. *The Kingdon Field Guide to African Mammals*. San Diego, CA: Academic Press.
Kistler, John M. 2006. *War Elephants*. Westport, CT: Praeger.
Langlois, John D., Jr. 1988. "The Hung-wu Reign, 1368–1398." In *The Cambridge History of China*. Vol. 7, *The Ming Dynasty, 1368–1644, Part I*, ed. Frederick W. Mote and Denis Twitchett, 107–81. Cambridge: Cambridge University Press.
Langman, Vaughan A. 1977. "Cow-Calf Relationships in Giraffe (*Giraffa camelopardalis giraffa*)." *Zeitschrift für Tierpsychologie* 43: 264–86.
———. 1982. "Giraffe Youngsters Need a Little Bit of Maternal Love." *Smithsonian* (January): 94–103.
Lankester, E. Ray. 1903. "On Okapia, a New Genus of *Giraffidae*, from Central Africa." *Transactions of the Zoological Society of London* 16: 279–315.
Laufer, Berthold. 1928. *The Giraffe in History and Art*. Chicago: Field Museum of Natural History.
Leuthold, Barbara M. 1979. "Social Organization and Behavior of Giraffe in Tsavo East National Park." *African Journal of Ecology* 17: 19–34.
Le Vaillant, François. 1796. *New Travels into the Interior Parts of Africa by Way of the Cape of Good Hope in the Years 1783, 84, and 85*. Vol. 2. London: G. G. and J. Robinson.
Lewis-Williams, David. 1983. *The Rock Art of Southern Africa*. Cambridge: Cambridge University Press.
———. 2002. *The Mind in the Cave: Consciousness and the Origins of Art*. London: Thames & Hudson.
Lewis-Williams, David, and David Pearce. 2005. *Inside the Neolithic Mind: Consciousness, Cosmos, and the Realm of the Gods*. London: Thames & Hudson.
Ma Huan. (1433, 1970) 1997. *Ying-tai Sheng-lan: The Overall Survey of the Ocean's Shores*. Trans. J. V. G. Mills. Bangkok: White Lotus.
McComb, Karen, Lucy Baker, and Cynthia Moss. 2006. "African Elephants Show High Levels of Interest in the Skulls and Ivory of their Own Species." *Biological Letters* 2 (October): 26–28.
Mejia, Carlos. 1971–72. "Giraffe Behaviour." *Serengeti Research Institute Annual Report*, 39.
Milius, Susan. 2003. "Beast Buddies: Do Animals Have Friends?" *Science News*, November 1, http://findarticles.com/p/articles/mi_m1200/is_18_164/ai_110737268/.
Mitchell, G., and J. D. Skinner. 2003. "On the Origin, Evolution and Phylogeny of Giraffes *Giraffa camelopardalis*." *Transactions of the Royal Society of South Africa* 58 (1): 51–73.
Moss, Cynthia. (1975) 1982. *Portraits in the Wild: Animal Behavior in East Africa*. Chicago: University of Chicago Press.
Mossop, E. E. 1947. *The Journals of Brink and Rhenius*. Cape Town: The Van Riebeeck Society.
Muller, Zoe. 2010. "The Curious Incident of the Giraffe in the Night Time." *Giraffa* 4 (1): 20–23.
Oppian. 1943. *Cynegetica*. Trans. A. W. Mair. Cambridge, MA: Harvard University Press.
Owen, Richard. 1839. "Notes of Giraffe Birth." *Proceedings of the Zoological Society of London*, 108–109.
Pager, Shirley-Ann. n.d. *A Walk through Prehistoric Twyfelfontein*. Windhoek: Typoprint.
Perry, John C. 2008. "Imperial China and the Sea." In *Asia Looks Seaward: Power and Maritime Strategy*, ed. Toshi Yoshihara and James R. Holmes, 17–31. Westport, CT: Praeger Security International.
Pournelle, George H. 1955. "Notes on the Reproduction of a Baringo Giraffe." *Journal of Mammalogy* 36 (4): 574.
Pratt, David M., and Virginia H. Anderson. 1979. "Giraffe Cow-Calf Relationships and Social Development of the Calf in the Serengeti." *Zeitschrift für Tierpsychologie* 51: 233–51.
———. 1982. "Population, Distribution, and Behaviour of Giraffe in the Aruha National Park, Tanzania." *Journal of Natural History* 16: 481–89.
———. 1985. "Giraffe Social Behavior." *Journal of Natural History* 19: 771–81.
Puxley, Frank Lavallin. 1929. *In African Game Tracks: Wanderings with a Rifle through Eastern Africa*. London: H. F. and G. Witherby.

Rice, E. E. 1983. *The Grand Procession of Ptolemy Philadelphus*. Oxford: Oxford University Press.

Roberts, Guy. 1969. "And Giraffe." *African Wildlife* 23: 171.

Roosevelt, Theodore. 1910. *African Game Trails: An Account of the African Wanderings of an American Hunter-Naturalist*. 2 vols. New York: Charles Scribner's Sons.

Savoy, James C. 1966. "Breeding and Hand-Rearing of the Giraffe *Giraffa camelopardalis* at Columbus Zoo." *International Zoo Yearbook* 6: 202–4.

Schiff, Stacy. 2011. *Cleopatra: A Life*. Boston: Little, Brown.

Schneemann, M., R. Cathomas, S. T. Laidlaw, A. M. El Nahas, R. D. G. Theakston, and D. A. Warrell. 2004. "Life-Threatening Envenoming by the Saharan Horned Viper *(Cerastes cerastes)* Causing Micro-angiopathic Haemolysis, Coagulopathy and Acute Renal Failure: Clinical Cases and Review." *QJM: An International Journal of Medicine* 97 (11): 717–27.

Selous, Frederick Courtenay. 1907. *A Hunter's Wanderings in Africa*. London: Macmillan.

———. 1908. *African Nature Notes and Reminiscences*. London: Macmillan.

Sherr, Lynn. 1997. *Tall Blondes: A Book about Giraffes*. Kansas City, MO: Andrews McMeel.

Silk, Joan B. 2002. "Using the 'F'-Word in Primatology." *Behaviour* 139 (2/3): 421–46.

Sinclair, John D. 1939. *Dante's* Inferno*: Italian Text with English Translation and Comment*. Oxford: Oxford University Press.

"Solitary Giraffe Gives Birth." 1966. *African Wild Life*, no. 20: 128.

Spinage, C. A. 1968. *The Book of the Giraffe*. Boston: Houghton Mifflin.

Stanley, Henry Morton. 1890. *In Darkest Africa: or, The Quest, Rescue and Retreat of Emin, Governor of Equatoria*. Vol. 2. New York: Charles Scribners' Sons.

Steinemann, Paul. (1955) 1965. *Cubs, Calves and Kangaroos*. London: Elek.

Stephan, Sol A. 1925. "Forty Years' Experience With Giraffes in Captivity." *Parks and Recreation* 9: 61–63.

Stiles, Dan. 2010. "The Okapi Wildlife Reserve: Is There Hope for the Future?" *Swara* 4: 47–49.

Strabo. 1930. *The Geography of Strabo*. Vol. 7. Trans. Horace Leonard Jones. New York: G. P. Putnam's Sons.

Street, Philip. 1956. *The London Zoo*. London: Odhams Press.

Tarou, Loraine R., Meredith J. Bashaw, and Terry L. Maple. 2000. "Social Attachment in Giraffe: Response to Social Separation." *Zoo Biology* 19: 41–51.

Thomas, Elizabeth M. (1958) 1989. *The Harmless People*. New York: Random House.

———. 2006. *The Old Way: A Story of the First People*. New York: Farrar, Straus and Giroux.

Thomas, O. 1894. "A Giraffe from Somaliland." *Proceedings of the Zoological Society of London*, 135–36.

Topsell, Edward. (1607) 1658. *The History of Four-Footed Beasts and Serpents*. London: G. Sawbridge.

Tsai, Shih-shan Henry. 1996. *The Eunuchs in the Ming Dynasty*. Albany: State University of New York Press.

Van der Jeugd, H. P., and H. H. T. Prins. 2000. "Movements and Group Structure of Giraffe (*Giraffa camelopardalis*) in Lake Manyara National Park, Tanzania." *Journal of Zoology, London*, 251: 15–21.

Van Sittert, Sybrand, John D. Skinner, and Graham Mitchell. 2010. "From Fetus to Adult—An Allometric Analysis of the Giraffe Vertical Column." *Journal of Experimental Zoology*, Part B, *Molecular and Developmental Evolution* 314B: 469–79.

Warren, James V. 1974. "The Physiology of the Giraffe." *Scientific American* 231 (5): 96–105.

Watson, Lyall. 1999. *Jacobson's Organ and the Remarkable Nature of Smell*. London: Penguin.

———. 2002. *Elephantoms: Tracking the Elephant*. New York: W. W. Norton.

White, T. H., ed. and trans. 1954. *The Book of Beasts*. New York: G. P. Putnam's Sons.

Williams, C. A. S. (1931) 1976. *Outlines of Chinese Symbolism and Art Motives: An Alphabetical Compendium of Antique Legends and Beliefs, as Reflected in the Manners and Customs of the Chinese*. 3rd ed. New York: Dover.

INDEX

H

I

J

K

L

M

N

O

DESIGN AND COMPOSITION: Nicole Hayward

TEXT: 9.5/14.5 Scala

DISPLAY: Interstate

PREPRESS: Embassy Graphics

INDEX: Alexander Trotter

PRINTING AND BINDING: QuaLibre